MODERN
PHYSICS
The Scenic Route

MODERN PHYSICS
The Scenic Route

Leo Bellantoni

Fermi National Accelerator Laboratory, USA

World Scientific

NEW JERSEY · LONDON · SINGAPORE · BEIJING · SHANGHAI · HONG KONG · TAIPEI · CHENNAI · TOKYO

Published by

World Scientific Publishing Co. Pte. Ltd.

5 Toh Tuck Link, Singapore 596224

USA office: 27 Warren Street, Suite 401-402, Hackensack, NJ 07601

UK office: 57 Shelton Street, Covent Garden, London WC2H 9HE

Library of Congress Cataloging-in-Publication Data
Names: Bellantoni, Leo, author.
Title: Modern physics : the scenic route / Leo Bellantoni,
    Fermi National Accelerator Laboratory, USA.
Description: New Jersey : World Scientific, [2022] | Includes bibliographical references and index.
Identifiers: LCCN 2021050500 (print) | LCCN 2021050501 (ebook) |
    ISBN 9789811242205 (hardcover) | ISBN 9789811243172 (paperback) |
    ISBN 9789811242212 (ebook for institutions) | ISBN 9789811242229 (ebook for individuals)
Subjects: LCSH: Physics.
Classification: LCC QC21.3 .B455 2022 (print) | LCC QC21.3 (ebook) |
    DDC 530--dc23/eng/20211201
LC record available at https://lccn.loc.gov/2021050500
LC ebook record available at https://lccn.loc.gov/2021050501

British Library Cataloguing-in-Publication Data
A catalogue record for this book is available from the British Library.

For any available supplementary material, please visit
https://www.worldscientific.com/worldscibooks/10.1142/12416#t=suppl

Typeset by Stallion Press
Email: enquiries@stallionpress.com

Printed in Singapore

Contents

Preface   ix

Acknowledgments   xiii

1. Symmetry   1

2. Mathematical Symmetries and Newton   5

3. A Symmetry That Is Not   9

4. Groups   17

5. Generators   23

6. Noether's Theorem   25

7. The Quantum Mechanical Robert Frost   31

8. The Central Procedure of Quantum Mechanics   35

9. Your First Quantum Calculation   39

10. Your First Quantum Experiment   43

11. What Heisenberg Didn't Know   47

12. Gauge Invariance   53

13. Where Do the Quanta Come From?   57

14. The Quest for Meaning: Particles and Waves   61

15. The Logos   65

16. Mental Waves   69

17. Einstein, Podolsky and Rosen   73

18. About Spin   79

19. Bell's Theorem: Setting up the Equipment   83

20. Bell's Theorem: Taking the Data   87

21. I Do Not Like It   91

22. If You Do Not Know Who Minkowski Was,
    What are You Doing in His Space?   93

23. Rotational Symmetries and Matrices   97

24. The Sort-of Rotation   101

25. 299,792,548 Meters per Second — and No More! 107

26. Going Slower by Going Faster   109

27. The Twins   111

28. Momentum in Minkowski Space   117

29. Why E Is In Fact mc^2 119

30. Antimatter   123

31. Your First Nuclear Physics Theory: Protons and Neutrons   127

32. Your First Nuclear Physics Theory: Symmetry   131

33. $SU(2)$: A Matrix Group   135

34. Your First Nuclear Physics Theory: Pions   139

35. Your First Particle Physics Theory: The Λ 143

36. Your First Particle Physics Theory: Strange Mesons   147

37. The Eightfold Way and Quarks   149

38. Another Symmetry that Is Not   155

39. γ, W, Z, and H 159

40. Bra-Kets   163

41. Two Fermions in a Pod   167

42. The Back of the Book   171

Index   189

Preface

"Wish I'd seen that sooner!"

There is by now a fairly conventional way to present modern physics at the undergraduate level, and that approach does have a lot to be said for it. In particular, a student comes out of a set of conventional courses with the ability to solve a collection of problems that are very similar to ones that are likely to need to be solved later on in another science or engineering class or job. Of course, those problem-solving skills also simplify the educator's task of assessing the student's nominal intellectual progress. This book exists though because we found, rather later in in our own personal educational trajectories, different approaches to some of the topics of an introductory modern physics class that are really illuminating as well as just plain nifty. We wish we had seen these different approaches a bit earlier on in the learning process. They do not make solving problems any easier, but they give us a whole new insight into how nature works.

While not forgoing the historical approach that is common in many modern physics courses (in fact we try to provide citations to the original works we discuss), we will try a more axiomatic approach to central concepts. In particular, we will place one of the more important ideas of modern physics, the idea of symmetry, front and center.

Unlike conventional texts, this book will also take some time to look over the fence that stands at the boundaries of what is scientifically known. For example, what is reality, really? So we will discuss the paper of Einstein, Podolsky and Rosen and their argument that quantum mechanics is not a complete description of reality, and Bell's argument that it is.

We will tend to focus on applications in nuclear and particle physics and not attempt a broad survey of all of the many fascinating subfields of physical science. Partly that is because symmetry plays such an important role in these subfields, partly because we personally find these subfields fascinating, and (as a result) partly because these are the subfields we know best!

This is not a conventional textbook, although it can serve well alongside such a book. It also will work on its own for readers possessing a certain level of mathematical skills. There is no calculus in here. There are complex numbers and matrices, which are covered in many pre-calculus courses. There are also groups, which all of us learn in some form in elementary school; only the names are changed to help us focus on specific ideas. We will need vectors, scalars, and the projections of vectors onto the axes of coordinate systems.

The other things that the reader will need are some of the basic ideas of physics. Things like the fact that momentum is a conserved vector quantity, that kinetic energy is $1/2mv^2$, that E is mc^2, the existence of isotopes, and so on. Nothing that you would not find in most high school advanced physics classes.

Do be warned though. Although most of the individual concepts here are routinely shown to high school students in advanced classes, the things we will do with them are not. A typical high school problem might use one or two concepts in a two or three step solution. Here we are doing much more complicated things, of many more steps in some places. So while our bricks are available at our high school, and some of the buildings we are constructing might be buildings on a high school campus, many of the buildings we are constructing are more likely to be found on a college campus. For the first time reader, the original papers that are cited are likely to be too difficult to follow; these citations are a resource that should be left for a later reading.

With this set of intellectual tools, we can explain why energy is conserved, how quantum mechanics works, why $E = mc^2$, understand antimatter, and do a little nuclear physics. We can categorize a bunch of subatomic particles, learn a fair amount about quarks, and understand why the Higgs boson is such a big deal. And, although it is not really our primary topic, we can fit a few paragraphs in about quantum computing.

The answers to all the exercises are in The Back of the Book. Some of the exercises are quick checks of understanding; some are really just derivations that were too long to fit into the main text.

Most importantly, we hope you enjoy these nifty ways to think about the basic results of relativity, quantum, nuclear and particle physics.

Acknowledgments

I want to thank very much Leah Hesla, science writer at Fermilab, whose careful reading and editing have made a clear message out of an unclear mess; it is her contribution that puts this book in the first person plural. I also wish to thank Don Lincoln for constructive feedback at an earlier stage of even greater messiness. Thanks also to Monica Noether, Zoe Steele, and Melissa Ma for being part of the journey; and to Chris Davis and Soh Yong Qi for their patient and generous support.

Chapter 1

Symmetry

Beginning around 1907, physicists began to realize that the idea of symmetry is particularly powerful in discerning the laws of nature. In the following century, it has become one of, if not the, most important idea in physics.

It was in that year that Einstein realized that if he were in a freely falling elevator, and dropped some object, it would not move away from him. That is exactly the same situation as if he were in outer space, away from any planet or star that would exert a gravitational force on him. The two situations are entirely equivalent and that is why this observation is called the equivalence principle; it did not necessarily strike Einstein at first that he was talking about a symmetry.

To be honest, it really does not look like a symmetry at all, in the way we normally use the word. We usually use *symmetric* to describe a property of a visual image. What we will need to do here is to be precise about what the word symmetric means. We need to rephrase our intuitive idea of symmetric into something a little more mathematically amenable.

Something is symmetric if we can

(1) *Say something describing what we observe.*
(2) *Transform what we are looking at in some particular way.*
(3) *Ask if we can still say the same thing you said in step 1. If so, we have symmetry.*

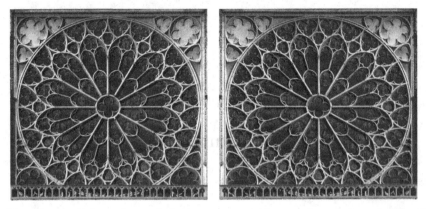

Figure 1.1: *Left*, a photo of the large rose window on the southern façade of Notre Dame de Paris cathedral. *Right*, the same photograph but flipped, left to right.

In ascertaining the symmetry in Figure 1.1, for step 1, the initial statement might be "Twelve large petal-shaped windows, each containing three smaller petal shapes and a four-lobed clover shape". To carry out step 2, to transform it, we flip the shape left to right, or form a mirror image around the vertical axis. In step 3, the concluding statement, "Twelve large petal-shaped windows, each containing three smaller petal shapes and a four-lobed clover shape" is exactly the same as the initial statement. Because the final and initial statement are the same, the image has this mirror symmetry.

Well, not exactly. If you look carefully, you can see a few very minor ways in which the left and right sides are slightly different. This symmetry is broken, but only a little broken.

Various parts of Figure 1.1 possess additional degrees of symmetry. The twelve petal-shaped structure inside the largest circle is symmetric not just for mirror flips around the vertical axis, but in fact around 24 different axes through the center of the circle. It also is symmetric with rotations; if the central circular area is rotated by any multiple of 30°, the initial and final statements will be the same.

Terminology: we say that this circular area is symmetric under rotations of 30° and mirror flips around 24 different directions; or we might say that it is symmetric with respect to these transformations, or that it is invariant under those or with respect to those transformations. If the symmetry does not exist, we say that it is violated or broken.

Here is how the idea of symmetry is applied in physics: the initial statement is a possible law of physics and the transformation is some sort of simple thing that should have no bearing on the physical situation or the possible law of physics. If the initial statement is no longer true after the transformation — if the symmetry is not there — then (most likely) our possible law is wrong; the thing that we thought might be a law of physics is not a viable law after all. Let us start by giving an example of symmetry in a purely mathematical object and then an example of symmetry in a law of physics.

Chapter 2

Mathematical Symmetries and Newton

The mathematical object we will use to demonstrate a mathematical symmetry is

$$\left(\frac{x-7}{5}\right)^2 + \left(\frac{y-3}{2}\right)^2 = 1. \tag{1}$$

This is the equation for an ellipse; it is centered at $x = 7$, $y = 3$. The height of the ellipse, from top to bottom, happens to be 4, and the width is 10.

The property of the ellipse that we will use is the area. The area of an ellipse is πab, where a is one half the height and b is one half the width. The area of this ellipse is $A = 10\pi$; that is our original statement — our step 1.

The transformation — step 2 — is to slide, or translate, the center of the ellipse over to some other place. Let us move the center to $x = 0$, $y = 12$. Then the equation is

$$\left(\frac{x-0}{5}\right)^2 + \left(\frac{y-12}{2}\right)^2 = 1. \tag{2}$$

The height of the ellipse, from top to bottom, is still 10 and the width is still 4. The area is still $A = 10\pi$; that is our statement after the transformation. Step 3 produces the same statement as the statement before the transformation. The equation for the ellipse, and the ellipse itself, which as a geometric object is a kind of mathematical object, has the property that the area is invariant with respect to the translation.

The ellipse's area has other symmetries. If x and y are swapped, which is a mirror reflection around the line $y = x$, the area will still not change.

There is another property of the equation that is symmetric and that is its form. By that, we mean the type of equation it is, and what is in it. The form of these ellipse equations, in English, is as follows. First, take x and subtract a number from it; divide that difference by another number and then take the square. Second, do that again but using y rather than x. Third, add those two together and set them equal to 1.

Both before and after the transformation, before and after sliding the center of the ellipse around, the equation has the same form. The form of the equation is symmetric with respect to translations.

Now for an example of a symmetry in a law of physics, to wit, Newton's 2nd law of motion. When a force F is applied to an object of mass m, then that object will have some acceleration a; and furthermore, $F = ma$. That is just for one dimensional motion, say, for the object moving left and right. Call that the x axis. The force in the x direction determines the acceleration in the x direction: $F_x = ma_x$. Is Newton's 2nd law symmetric when it is flipped from left to right?

After the mirror transformation, the force changes sign. Suppose that the force is pushing towards my right. From the point of view of the "man in the mirror", it is pushing towards his left. If $F_x > 0$, the mirrored force is negative; we replace F_x with $-F_x$. Similarly, the mirrored acceleration points in the opposing direction, and we replace a_x with $-a_x$. The mass however stays the same. The mass is a number to describe how much inertia the object has, and it does not depend on the direction in which the force is applied. It is a scalar, not a vector.

So, replace F_x with $-F_x$ and replace a_x with $-a_x$, and we have $-F_x = m(-a_x)$, which is just the same equation as the original one, $F_x = ma_x$. Newton's 2nd law in one dimension is symmetric with respect to the mirror transformation.

Had Newton written $F_x = ma_x^2$, well, that just would not have made him famous. Newton was trying to describe the motion of rather ordinary objects — apples, clock pendula, fluids, and such. Broadly speaking, these motions look the same in the mirror as they do directly. Newton's second law of motion has to have to have the mirror symmetry, but $F_x = ma_x^2$ does not have that symmetry. Replacing again F_x with $-F_x$. and a_x with $-a_x$ leaves $F_x = -ma_x^2$, which is not the same as the original. Therefore, $F_x = ma_x^2$ cannot be right.

Chapter 3

A Symmetry That Is Not

In the 1950s, physicists discovered a symmetry that was not, to our great astonishment, there. That led to a great step forward in our understanding of a whole category of interactions between the elementary particles.

To start with, we must clarify the statement "Broadly speaking, these motions look the same in the mirror as they do directly."

To be precise, vectors like forces $\vec{F}$, distances $\overrightarrow{\Delta x}$ or accelerations $\vec{a}$ do indeed change sign when the image is flipped as if in a mirror. There are other quantities that do not. For starters, the product of two numbers that change sign will not change sign. Work, that is $W = \vec{F} \cdot \overrightarrow{\Delta x}$, will be unchanged: $(-\vec{F}) \cdot (-\overrightarrow{\Delta x}) = \vec{F} \cdot \overrightarrow{\Delta x}$.

Quantities that change sign with the mirror transformation are said to have odd parity and those that do not are said to have even parity. Notice that the whole idea of saying that something has even or odd parity depends on the mirror symmetry being good, on the laws of physics being the same when flipped in the manner of Figure 1.1.

Now there are two ideas that come with parity. The first is that it is conserved; it does not change with time. The vector $\vec{F}$, which has odd parity today, will have odd parity tomorrow. The second is the way parities combine:

$$
\begin{aligned}
\text{even with even} &= \text{even} \\
\text{even with odd} &= \text{odd} \\
\text{odd with even} &= \text{odd} \\
\text{odd with odd} &= \text{even}
\end{aligned}
\tag{3}
$$

We will say more about this later, in Groups.

In 1950, physicists did not have a good idea of what a pion actually is. They had Yukawa's theory of 1935 (see Your First Nuclear Physics Theory: Pions) but that is really about what a pion does; it does not really explain what it is made from or what is in it. If you could flip a pion left to right, would it have even parity? Or odd? What kind of a mathematical object should be used to describe the inside of a pion?

Experiments done in 1951 and 1954 showed that pions have odd parity.[1] Earlier, in 1949, there had been the discovery of a charged particle with a mass about half that of the proton and decaying into three pions, and therefore, you might think,[2] of odd parity.[3] This particle was called the tau meson, τ, although that name is no longer used. At about the same time, researchers were also finding another charged particle, sometimes called a K meson, which is the name used now, but also called the θ meson. The θ also had a mass about half that of the pion, but it decayed into two pions, not three, and so must have even parity. These were a disturbing set of results. By 1956 it was clear that the θ and the τ had not just the same masses, but also the same lifetimes, which does make it seem like they are indeed the same particle. But this should not be possible; the θ and the τ have different parities and so cannot be the same particle, even though they have the same mass, lifetime and charge.

The beginning of learning is to realize what you do not know and thought you did. What did we get wrong here? Parity conservation had been used for 30 years by the mid-1950s and had always worked just fine. Why did it suddenly fail?

---

[1]W. K. H. Panofsky, R. L. Aamodt and J. Hadley, "The Gamma-Ray Spectrum Resulting from Capture of Negative π-Mesons in Hydrogen and Deuterium", *Phys. Rev.*, 81, 4, 565–574 (1951); W. Chinowsky and J. Steinberger, "Absorption of Negative Pions in Deuterium: Parity of the Pion", *Phys. Rev.*, 95, 6, 1561–1564 (1954).

[2]There is a technical detail. The produced pions must not, and indeed do not, have any orbital angular momentum, that is, the motion of their centers of mass must not have any net overall rotation.

[3]R. Brown, U. Camerini, *et al.* "Observations with Electron-Sensitive Plates Exposed to Cosmic Radiation", *Nature*, 163, 82–87 (1949).

Decays happen because of forces. For example, calcium carbonate will, under certain conditions, decay into calcium oxide and carbon dioxide: $CaCO_3 \rightarrow CaO + CO_2$. That is a chemical process; what happens is that electrons in the molecule move from one place to another. Before the move the electrons are in a place where they bind oxygen to carbon; afterward, they are in a different location where they bond the calcium atom to one of the oxygen atoms. The point is that the forces that cause the electrons to move are electrical. When the 143 neutrons and 92 protons that form a ^{235}U nucleus rearrange themselves to make a ^{231}Th nucleus and a ^{4}He nucleus, that is when a ^{235}U emits an alpha particle, those neutrons and protons move around in response to the force that holds them together. That is the strong nuclear force; we will just call it the strong force.

In 1956, T. D. Lee and C. N. Yang[4] realized that all the cases where physicists knew that parity is conserved were cases where the forces were either electromagnetic or strong forces. However, the force involved in θ and τ meson decays are neither of these. The force involved in these decays is the weak nuclear force. That is the force that causes β decay, among other things; we will just call it the weak force. For over 20 years, people had studied the weak force, "knowing" that it conserves parity. Except that they did not know that, any more than Urban VIII knew that the earth does not move.

After realizing that parity might not be conserved and that this would solve the θ–τ puzzle, Lee and Yang proposed an experiment using ^{60}Co to check if the mirror symmetry does indeed exist for weak force decays. That experiment was done[5] in 1956 by C. S. Wu and her colleagues.

In a cobalt nucleus with 33 neutrons, one of the neutrons, n, will decay into a proton, p, an electron, e^-, and an antineutrino, $\bar{\nu}$. Antineutrinos (and neutrinos) are rather like electrons without

---

[4]T. D. Lee and C. N. Yang, "Question of Parity Conservation in Weak Interactions", *Phys. Rev.*, 104, 254–258 (1956).

[5]C. S. Wu, E. Ambler, R. W. Hayward, D. D. Hoppes and R. P. Hudson, "Experimental Test of Parity Conservation in Beta Decay", *Phys. Rev.*, 105, 1413–1415 (1957). As best we can tell, it may well have been C. S. Wu who suggested the experiment to Lee and Yang.

an electrical charge. The newly created proton stays in the nucleus, so the result is a ^{60}Ni nucleus, nickel of course being the element that has 28 protons in its nucleus. The antineutrino and the electron head off in opposite directions with considerable momentum, but the neutron and proton are basically at rest. The neutron and proton stay put because of their considerable mass, and are trapped inside the nucleus.

All these particles have a property called spin. Spin is angular momentum, built into the particle, like a permanently spinning top. Just as a subatomic particle has a specific mass and electric charge, it also has a specific amount of angular momentum. We will discuss spin in more detail later on, in About Spin. All the particles in this decay process have the same amount of spin, and we can get by with thinking of each particle as spinning to the left or to the right. Like charge or momentum, angular momentum is conserved, and in this case the spin, the intrinsic angular momentum of the particles is the only kind of angular momentum involved. There is no angular momentum from one particle orbiting around another. So, the total spin of all the particles is the same before and after the decay process. Since in this case we start with one particle and we finish with three particles, we must create two particles with opposite and hence cancelling spins.

Figure 3.1 shows a possible decay. The initial state, a neutron with a certain spin, is shown on the left. We will call this a

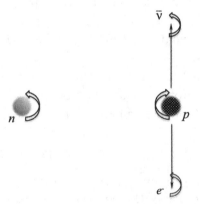

Figure 3.1: At left, a neutron at rest, before decay. At right, the products of the decay: a proton at rest, an electron and an antineutrino recoiling in opposite directions, all with specific spins.

right-handed spin. On the right, the final state. The spin of the electron and proton are equal and opposite; the electron has a right-handed spin and the proton has a left-handed spin. These two cancel each other, so the antineutrino must have the right-handed spin, so the total spin is that of the initial neutron.

Now, what appears in the mirror? What happens is the spins, in our convention, flip around. A particle with a right-handed spin, such as the neutron in Figure 3.1, will appear with a right handed spin as seen by the "man in the mirror", which is a left-handed spin for those of us not in the mirror, as in Figure 3.2. Stand in front of a mirror with a ball or something like that and rotate it slowly to help visualize this.[6] The spins of the decay products, the electron, proton, and antineutrino should all flip, too.

Here is the surprising thing, which made this experiment so famous: the transition from the left to the right side of Figure 3.2 does not happen. That is what C. S. Wu and her colleagues found: the mirror symmetry is just not there for decays involving the

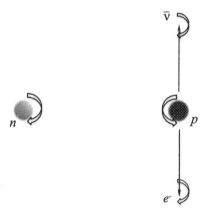

Figure 3.2: The mirror image of Figure 3.1, which astonishingly, does not happen!

---

[6]The spin of a particle, like all angular momenta, is an even parity quantity. It is the product of a distance, which is odd parity, with a velocity, which is also odd partity. The "man in the mirror" sees the object spinning in the same direction as the "man before the mirror". Figure 3.2 is from the point of view of the "man before the mirror".

weak force! A $\bar{\nu}$ must have the right-handed spin as in Figure 3.1 and never[7] has the left-handed spin as in Figure 3.2. The laws of the universe, or these particular laws at least, regard turning to the left as fundamentally different from turning to the right. That is a peculiar thing to have discovered!

It turns out to be more than just peculiar. The fact that parity is not conserved was one of a series of discoveries about the weak force. Particles like electrons, neutrons, protons, and neutrinos have antimatter versions. We will be able to say more about antimatter later, after discussing relativity. The electromagnetic force acts on antimatter in the same way as it acts on matter. The strong force, likewise. But the weak force is totally different for matter and antimatter.

But wait, there is more. The difference in the weak force between matter and antimatter is, to very good approximation, cancelled by the nonexistence of the mirror symmetry. In Figure 3.1, there is an antineutrino with right-handed spin, and that particle exists. In Figure 3.2, there is an antineutrino with left-handed spin, and that particle does not exist. What if it were a neutrino, rather than the antimatter form, the antineutrino? For neutrinos, the left-handed spin exists and the right-handed spin does not exist. Use P to indicate the process of applying the mirror symmetry, and C to indicate the operation of replacing every particle with its antiparticle. Then symmetry with the P operation is broken in weak force interactions. Also, the symmetry with the C operation is broken in weak force interactions. But if we apply P and C together, then there is a symmetry; CP is an unbroken symmetry.

Oh, but wait. It turns out that CP is only approximately unbroken. In 1964, Christenson, Cronin, Fitch and Turlay discovered[8] that

---

[7]Well, as the captain of the *Pinafore* said, hardly ever. In the years since, we have learned that neutrinos must have a very tiny amount of mass. That means that in principle at least, one could have a $\bar{\nu}$ with a spin as in Figure 3.2. But such a thing has never actually been seen in a laboratory.

[8]J. H. Christenson, J. W. Cronin, V. L. Fitch and R. Turlay, "Evidence for the 2π Decay of the K_2^0 Meson", *Phys. Rev. Lett.*, 13, 138–140 (1957).

about 0.2% of the time, a certain particle, the K_2^0 meson, decays in a way that does not conserve *CP*.

Now there is another odd thing out there. The universe is made almost entirely of matter; there is very little antimatter seen in our telescopes. That asymmetry could be explained if certain things, known as the Sakharov conditions, are true. One of the Sakharov conditions is that *CP* is a broken symmetry. So the symmetry that is not, the mirror symmetry, might be connected somehow to the existence of our universe, being created out of matter and not antimatter. Science has not proven that the connection exists, but it is a possibility.

Chapter 4

Groups

The mathematical object that is most useful in describing symmetries is the group. Let us start with some examples.[1]

Start with a square piece of some transparent material such as glass. Rotate it counterclockwise through a right angle so that the corner in the upper right is now in the upper left. Voila, it is still a square. This is just the same sort of thing that we did with the windows of Figure 1.1. Squares are symmetric with rotations through $90°$. They are also symmetric with rotations of $180°$, $270°$ and $360°$. (These angles are all given for counterclockwise rotations). They are also symmetric with a rotation through $0°$; that is the operation that consists of doing nothing at all.

Moreover, two rotations of $90°$ is a rotation of $180°$. We write these rotations as R_{90}, R_{180} and so on, and we write consecutive rotations as if they were multiplications, using the dot symbol ($\cdot$). So $R_{180} = R_{90} \cdot R_{90}$. The fact that a $90°$ rotation followed by a $270°$ rotation is a $360°$ rotation can be written as $R_{360} = R_{270} \cdot R_{90}$. The operations are written right to left;[2] $R_{270} \cdot R_{90}$ represents a $90°$ rotation followed by a $270°$ rotation, not a $270°$ rotation followed by a $90°$ rotation.

In simple terms, a group is a nonempty set of things and an operation upon with any two of the things. In this case the set of

---

[1] This introduction follows that of J. A. Gallian, Contemporary Abstract Algebra (Boston, MA: Brooks/Cole, 2013), which is an excellent introduction to groups.
[2] This is intentionally akin to how matrix multiplication is done.

things is the set of rotations. The operation that can be done upon two rotations is to do one rotation and then the other. For the set and the operation to qualify as a group, it has to meet four conditions, which we will describe.

Turning back to our example, these rotation operations commute. For example, a 90° rotation followed by a 180° rotation is the same as a 180° rotation followed by a 90° rotation: $R_{180} \cdot R_{90} = R_{90} \cdot R_{180}$. When the elements of a group commute, the group is called Abelian, after the 19th century Norwegian mathematician Niels Abel.[3]

There are other symmetries to the square. There is the mirror symmetry, which is a complete flip about the vertical axis. Flips around the horizontal axis or across either of the two diagonals also preserve symmetry. Call these flips F_{90} (around the vertical axis), F_0 (around the horizontal axis), F_{45} (around the diagonal that cuts from the upper right to the lower left), and F_{135} (around the diagonal that cuts from the upper left to the lower right). The subscript here gives the axis of the flip. So $F_{90} \cdot F_{90} = R_0 = F_{135} \cdot F_{135}$, and so on. To keep track of combined flip and rotation operations, it helps to think of one corner of the square painted a particular color and a second corner painted a different color. Now $R_{90} \cdot F_0$ is F_{45}, but $F_0 \cdot R_{90}$ is F_{135}, so when flips are included in the group, the group is non-Abelian. The rotation subgroup is Abelian though.

Now we can write a multiplication table, also called a Cayley table. The multiplication tables that you learned in elementary school are just special cases of these. The first operation to be applied, the one written on the right, is in the header row, and the header column is the operation on the left, the second operation to be applied. The table is 8×8, because there are eight elements in the group: R_0, R_{90}, R_{180}, R_{270}, F_0, F_{45}, F_{90} and F_{135}. The rotation subgroup is the 4×4 part of the table in the upper left, and the flips form an Abelian subgroup in the 4×4 part of the table in the upper right.

Every location in Table 4.1 is filled with an element of the group, and the entire table can be filled in using only our eight elements.

---

[3]What is purple and commutes? An Abelian grape! Dweeb humor is really pathetic, no?

Table 4.1: The Cayley table for symmetries of the square. Entries are the value of rc, where r is the row, labeled at left, and c is the column, labeled at the top.

	R_0	R_{90}	R_{180}	R_{270}	F_0	F_{45}	F_{90}	F_{135}
R_0	R_0	R_{90}	R_{180}	R_{270}	F_0	F_{45}	F_{90}	F_{135}
R_{90}	R_{90}	R_{180}	R_{270}	R_0	F_{45}	F_{90}	F_{135}	F_0
R_{180}	R_{180}	R_{270}	R_0	R_{90}	F_{90}	F_{135}	F_0	F_{45}
R_{270}	R_{270}	R_0	R_{90}	R_{180}	F_{135}	F_0	F_{45}	F_{90}
F_0	F_0	F_{135}	F_{90}	F_{45}	R_0	R_{270}	R_{180}	R_{90}
F_{45}	F_{45}	F_0	F_{135}	F_{90}	R_{90}	R_0	R_{270}	R_{180}
F_{90}	F_{90}	F_{45}	F_0	F_{135}	R_{180}	R_{90}	R_0	R_{270}
F_{135}	F_{135}	F_{90}	F_{45}	F_0	R_{270}	R_{180}	R_{90}	R_0

No new rotations or flips were needed to fill in the table. This is the closure property; it is the first of the four requirements for a set and an operation to be a group.

The second requirement is that a group's operations must have the associative property. For every A, B, and C in the group, it must be true that $A \cdot (B \cdot C) = (A \cdot B) \cdot C$. For example, $F_{135} \cdot (F_{90} \cdot R_{270}) = (F_{135} \cdot F_{90}) \cdot R_{270}$. The former is, using Table 4.1, $F_{135} \cdot F_{135} = R_0$; the latter is $R_{90}R_{270} = R_0$.

The element R_0 has a special property; it does nothing whatsoever. For every other element of the group, E, the combined operation $E \cdot R_0 = R_0 \cdot E = E$. We call R_0 the identity element, and usually write it as $\mathbb{1}$. Every group must have an identity element; that is the third group condition. The identity element commutes with every element of the group, even if it is a nonAbelian group.

Exercise 1. Show there is only one identity element in any group.

The fourth and final part of the definition of a group is that each element of the set must have an inverse; so if A is in the group, then there is some B which is also in the group, so that $A \cdot B = B \cdot A = E$ is the identity element. The inverse of A is written as A^{-1}.

Exercise 2. Prove that every element of a group has only one inverse.

Exercise 3. Notice how every element of the group given by Table 4.1 appears exactly once in each row. Using the existence of an inverse for each element, prove that this is true for every group.

The symmetries of the square are a discrete group: it has a countable number of elements.[4] There are also continuous groups. An example of a continuous group which we use later is the rotational symmetries of a circle. It is possible to rotate a circle in a plane by an angle that is given by any real number. There is a parameter, call it φ, that identifies every particular element of the group by identifying the angle through which the circle is being rotated. Because φ is a real number, it can be changed continuously; φ can change smoothly, without visible steps. There are an infinite number of elements in the circle group.

Exercise 4. For the rotations of a circle around its center as it lies in a plane, identify the set, the associative binary operation, and the identity element. Show that the operation is closed and invertible.

Exercise 5. Why is not the set of all integers, with the operation of multiplication, a group?

Now we can see something special about the rules for combining parity in Equation (3). This is a set with two elements, "even" and "odd". We want an operation to combine them. The order cannot matter; an even parity particle and an odd parity particle must combine to the same parity as an odd parity particle with an even parity particle. So the group we are looking for must be Abelian. There is only one group with two elements, and it is indeed Abelian; its Cayley table is summarized in Equation (3). The group is named Z_2; it can also be described as the set of integers with the operation of addition modulo 2. Table 4.2, is the Cayley table for Z_2 with three different labels for the two elements. When discussing parity, we have used the labels "even" and "odd". Were we discussing the sign of the product of multiplying two real numbers, we would use the labels "+" and "−", as in "+3 times −2 is −6". Or we could use the labels

---

[4]This group is called D_4, the dihedral group of order 4.

Table 4.2: The Caley table for Z_2.

	even, + , or 0	odd, − , or 1
even, + , or 0	even, + , or 0	odd, − , or 1
odd, − , or 1	odd, − , or 1	even, + , or 0

0 and 1 emphasize that the operation is addition. But whatever the labels are, the Cayley table is the same:

There is of course the group Z_3, the set of integers with the operation of addition modulo 3, the group Z_4, and so on. These are called the cyclic groups. The rotation subgroup of the upper left of Table 4.1 is Z_4.

Exercise 6. Find an example of Z_{12}.

Chapter 5

Generators

An important idea from the world of group theory is the idea of the generator. Consider the group $\{R_0, R_{90}, R_{180}, R_{270}\}$, which is the group Z_4. The R_{90} element has the property that you can reach every element of the group by multiplying by it some number of times. To get to R_{270} for example, take $R_{90} \cdot R_{90} \cdot R_{90}$. Such an element is called a generator of the group. R_{180} is not a generator of this group; you can get only to R_0 and R_{180} by repeatedly multiplying R_0 by R_{180}. R_{270} is a generator, but it is also the inverse of R_{90}; R_{270} cycles through the elements of the group in the reverse order as R_{90}. So there can be more than one generator in a group.

By itself, R_{90} does not generate the entire group of symmetries of the square as shown in Table 4.1, but R_{90} together with one of the flips, say F_0, does. That is, every element of the entire group can be obtained by some sequence of 90° rotations and flips around the horizontal axis. We say that $\{R_{90}, F_0\}$ is a generating set for the group in Table 4.1. There are seven other combinations of either R_{90} or R_{270} with any of the four flipping operations, and those two are generating sets. But there will always be two elements in a complete set of generators for this group.

For a continuous group, such as the rotations of a circle, we will not give the precise definition of a generator. Generators for continuous groups are constructed using infinitesimal numbers, such as you may have met in studying calculus. However, generators here have the same basic function: every element of the group can be written as some combination of the generators. In the cases of interest for us, the number of elements in a complete set of generators will be the dimensionality of the group. For example, the circle group has only one generator. One can only rotate a circle in one direction, so to speak. However, the set of vectors $[x, y, z]$ in three dimensional space, with the operation of vector addition, is a continuous group with three generators. A sphere can be rotated around three different axes, and the group of rotations of a sphere has three generators.

Chapter 6

Noether's Theorem

Not enough people know about Emmy Noether.

In 1918, she proved two theorems[1] about symmetry that underlie much of the theoretical physics that has happened since then. Her first theorem, which involves the Lagrangian, is of tremendous importance. The Lagrangian is a mathematical expression that summarizes all the laws of physics that apply for a given situation. As an example, the Lagrangian for an object moving in free fall near the surface of the earth, neglecting air resistance is

$$L = \frac{1}{2}mv^2 - mgh \tag{4}$$

where m is the mass of the object, v is its speed, h is the height of the object, and $g = 9.8\,\text{m/sec}^2$, the acceleration from gravity.

The law of physics that you would get from the Lagrangian in Equation (4) is just $a = -g$, and from that you can derive that the height of the object at any specific time t is $h = h_0 + v_0 t - (g/2)t^2$, where h_0 and v_0 are the initial height and velocity of the object. To actually get to $a = -g$, from Equation (4) requires the use of a fairly non-trivial technique called the calculus of variations, which we will not try to get into. The key thing is the idea that if you know the Lagrangian for a given system, you can derive all the laws of physics for that situation.

---

[1]The theorems were actually presented by Felix Klein on 26 July of that year, as the idea that women could do mathematics was still ... novel.

Figure 6.1: Amalie Emmy Noether. Courtesy Noether family.

Lagrangians are not unique; if some specific Lagrangian L describes a system, and L is either multiplied by a constant m or added to a constant b, the new Lagrangian $mL + b$ summarizes the same laws of physics as the original one.

Noether's theorem applies in the case where the Lagrangian's form is invariant under a continuous transformation. Recall that the form of the equation of an ellipse does not change with respect to the transformation of sliding the center of the ellipse around in a plane. Noether's theorem applies in analogous cases. It does not apply for a discrete symmetry, such as the mirror symmetry or Z_2.

Noether's theorem says that in such cases, there is a conserved physical quantity. Not only does the theorem tell you that there is a conserved quantity, it also tells you how to find the conserved quantity. The math is a little advanced, and we will skip it. But there are some really astonishing results when you find that conserved quantity.

Let us work with the case of the Lagrangian in Equation (4). Suppose Drs. Grassi and Sarsi choose to reenact Galileo's famous experiment at the Leaning Tower of Pisa. As a first step, rewrite the Lagrangian in more detail. Use x, y, and z to give the position of a cannonball relative to some spatial origin, with the stipulation that z is the vertical direction, the direction of motion. Use t for the time relative to some specific moment $t = 0$. Then the Lagrangian for a single cannonball is

$$L = \frac{1}{2}m\left\{\left(\frac{\Delta x}{\Delta t}\right)^2 + \left(\frac{\Delta y}{\Delta t}\right)^2 + \left(\frac{\Delta z}{\Delta t}\right)^2\right\}^2 - mgz. \qquad (5)$$

The terms $\Delta x/\Delta t$, $\Delta y/\Delta t$ and $\Delta z/\Delta t$, are the velocities in the three directions; they are the rate at which the positions $[x, y, z]$ are changing. If cannonballs are released at a height of 50 m above the ground then, according to the laws of physics summarized in this Lagrangian, they will be 32.31 m above the ground after falling for 1.9 seconds, and 28.39 m above the ground after falling for 2.1 s. That makes the velocity $(28.39–32.31)/0.2 = -19.6\,\text{m/s}$ after 2 seconds; the negative sign is for the downward direction.

Now for the transformation. Drs. Grassi and Sarsi decide that they will take their data independently and compare their notes afterwards. Unbeknownst to Dr. Grassi, Dr. Sarsi takes all his height measurements relative to the bottom of the foundation of the building, which is a distance b below the ground. Dr. Grassi measured from the ground.

There is nothing wrong with that. The laws of physics must be the same wherever the $[x, y, z]$ coordinate system has its origin.[2] This is called the Principle of Relativity.

Dr. Grassi calculates the velocity just as we did, but Dr. Sarsi calculates differently. After 1.9 s, he records the cannonballs at height

---

[2]This is a simplification. It is possible to set up coordinate frames where objects that have no forces acting upon them accelerate; these are called non-inertial frames. They are not relevant for the discussion here.

$(32.31 + b)$ m; after 2.1 s, he records them at $(28.39 + b)$ m. Then he finds the velocity after 2 s to be $((28.39 + b) - (32.31 + b))/0.2 = -19.6$ m/s. That is, moving the zero point for the measurement a distance b in the vertical direction does not change the velocity:

$$\frac{\Delta z}{\Delta t} = \frac{(z_a + s) - (z_b + s)}{t_a - t_b} = \frac{z_a - z_b}{t_a - t_b}. \tag{6}$$

Not only is the velocity unchanged, the mgz term is effectively not changed. Drs. Grassi and Sarsi will indeed have different numbers for mgz, but they will differ by the constant mgb and Lagrangians that differ by a constant are really the same.

In terms of the three steps that define a symmetry:

(1) *Say something describing what we observe:* In this case, the initial statement is to assert that Equation (5) will describe the motion of the cannonballs.
(2) *Transform what we are looking at in some particular way:* The transformation is to shift the origin of the coordinate system in amount b.
(3) *Ask if we can still say the same thing we said in step 1. And if so, we have symmetry:* What we have worked out is that after changing the coordinate system, it is still true that Equation (5) will describe the motion of the cannonballs.

Actually, the coordinate system can be shifted in three directions, with three real numbers; one for the x, one for the y and one for the z directions. So actually this symmetry is described by the group of three dimensional vectors with the operation of addition. Then — we omit the math here — Noether's theorem tells you what the three conserved quantities are. They are the three components of the momentum of the cannonball in three dimensions, a vector.

What about a time offset? What about a Principle of Relativity for time, as well as for space? Dr. Sarsi, it turns out, has just flown in from Dublin and has forgotten to set his watch forward an hour. So all his time readings differ from Dr. Grassi's by $\tau = -1$ hr $= -3600$ s.

Still, the Lagrangian is unchanged:

$$\frac{\Delta z}{\Delta t} = \frac{z_a - z_b}{(t_a + \tau) - (t_b + \tau)} = \frac{z_a - z_b}{t_a - t_b}. \tag{7}$$

Here there is only one parameter; the symmetry is described by the set of real numbers (τ), with the operation of addition. The conserved quantity as calculated from Noether's theorem is energy!

We have learned two things. The first is that energy is to momentum as time is to space. We will see the same idea expressed in a different way when we get into space-time. The second is that conservation of energy and momentum can be seen as consequences of the Principle of Relativity. Now, figuring out why energy and momentum are conserved is A Big Deal! Which is why, really, more people should know about Emmy Noether.

Chapter 7

The Quantum Mechanical Robert Frost

It is common to think of quantum mechanics as the laws of motion for extremely small objects, such as electrons and atoms. Indeed, the first experiments that led to the development of quantum mechanics were all done with extremely small objects. But there is more to it than this. Quantum mechanics is needed to explain superfluidity, and superfluidity is not a microscopic phenomenon. Neutrino oscillations are fundamentally quantum mechanical in nature, and the distance scale for some neutrino oscillation experiments is the distance from the sun to Earth.

It is much better to think of quantum mechanics as the laws of motion for when you cannot tell distinguish between several alternatives. For example, there is no way to tell two electrons apart; you need to apply quantum mechanics to determine what happens when you have two of them in the same place. In situations where there are two or more alternative things that can happen, and you cannot tell which of them actually does happen, you need quantum mechanics. We give two specific examples.

Our first example[1] uses a device called an electron gun. If you take a wire made of tungsten and make it red hot in a vacuum,

---

[1] Our first example is from Feynman's exposition, presented at a somewhat more advanced level than here in R. P. Feynman and A. R. Hibbs, *Quantum Mechanics and Path Integrals* (New York: McGraw-Hill, 1965). See also Vol. 3 of R. P. Feynman, R. B. Leighton and M. Sands, *The Feynman Lectures on Physics* (Reading Massachusetts: Addison-Wesley, 1965).

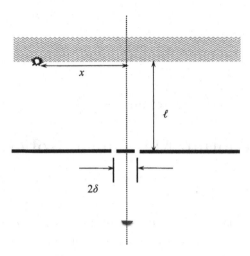

Figure 7.1: The prototypical quantum mechanical two-alternatives problem.

electrons can boil off of it. Put that hot wire near a piece of metal with a positive voltage on it and a little hole in it, and the boiled-off electrons will be pulled towards that metal and some of them will come out through the hole on the other side. This electron gun, as it is called, is shown as a little semicircle at the bottom of Figure 7.1. Now let those electrons fly through space for a while and head for a thin sheet of metal with two slits cut in it. That is shown as a horizontal line with two gaps for the slits. Call the spacing between the slits 2δ. The experiment works better when δ is small, and δ of $0.5\,\mu$m can be arranged fairly easily. The electrons that hit the metal go into it and are lost to us. The electrons that do not hit the metal have two options; they can fly through either the left or the right slit. Which slit the electron goes through cannot be determined; more on that later. That set of two indistinguishable options is what makes this a quantum mechanical problem. Then a detector, shown as a rectangle at the top of Figure 7.1, lights up at the place where an electron hits is placed some distance ℓ behind the two slits. The interesting thing to look at is the position on that detector, x.

Our second example involves matter and antimatter. The antimatter form of an electron is a positron; it is just like an electron but has a positive rather than a negative charge. There is another particle, the muon, which is just like an electron but for no obvious reason has more mass. And — no surprise — there is an antimatter form of the muon, which has the higher mass and is positively charged. If an electron and a positron collide, then they annihilate and make a bundle of pure energy. This energy can take two possible forms. It might be in the form of electric and magnetic fields, or it could be energy in the fields of the weak force. Regardless of the kind of energy that is formed, that energy can reappear as another pair of particle and antiparticle. Our second example is the case where the energy reappears as a muon and an antimuon. These two are produced going out from the collision point in some direction.

Instead of measuring a distance x as in the two-slit example, we measure the direction of the muon/antimuon pair. Again, the thing that makes it a quantum mechanical problem is that there are two possible alternatives, two possible series of events, for the system to get to the same result: the muon and antimuon could be formed from either the electromagnetic energy or the weak force energy. Notice that in this case, the entire process takes place at a single point in space; there are no distances like δ or ℓ in the problem.

These examples, where it is impossible to tell which of the alternatives actually occurred, are examples of interfering alternatives. If it possible to know which path the system took, there are exclusive alternatives. Suppose that, instead of considering an electron going through two slits, we consider a baseball going through two windows. In that case, it is quite obvious from the broken glass which alternative occurred, and quantum mechanics is not called for.

Similarly, there are two remarkably similar grey rubber doorstops over by the door there. They appear to have come out of the same mold. Quantum mechanics is not needed to explain anything about them because they are so large that they can readily be distinguished. There is one on the left and one on the right, and they are clearly two different doorstops. If they are weighed with enough accuracy and precision, some minor difference between the two will appear and serve to distinguish them. This is quite unlike the case of two electrons in close proximity to each other; every electron is the same, and it is impossible to distinguish between them.

Chapter 8

The Central Procedure
of Quantum Mechanics

When dealing with exclusive alternatives, the probabilities for each of the two alternatives to occur are real numbers that add. If the probability of the electron going through the slit on the left and landing at x (within some experimental precision δx)[1] is $P_L(x)$ and the probability for it to go through the slit on the right and landing at x is $P_R(x)$, then under the (incorrect, in this case) assumption of exclusive alternatives, the total probability of the electron appearing at x, the probability of outcome x, is

$$P(x) = P_L(x) + P_R(x). \tag{8}$$

That is, in essence, the nonquantum case. For the quantum case, where we must accommodate interfering alternatives, the procedure is:

(1) Replace the real numbers $P_L(x)$ and $P_R(x)$ with complex numbers $A_L(x)$ and $A_R(x)$, which are called amplitudes. We will not get into procedures for how to calculate amplitudes, but the sole input to the process is the Lagrangian, which summarizes the laws of physics that apply for any given situation.

---

[1]Strictly speaking, we should distinguish the probability density from the probability; the probability for the electron to appear between $x - \delta x/2$ and $x + \delta x/2$ is given by δx times the probability density. For our discussion it is sufficient to simply say "probability" even if in fact it is the probability density that we are discussing.

(2) The probability of outcome x for interfering alternatives is

$$P(x) = N\Phi(x)|A_L(x) + A_R(x)|^2. \qquad (9)$$

The factor $\Phi(x)$ is not something we want to spend a lot of time on. It is called the density of states and is to allow for the possibility that for simple geometric or counting reasons, there might be more than one way to have an outcome x. For example, if the two slit experiment is repeated with alpha particles going through two slits rather than electrons, the density of states for the alpha particle case will be half the density of states for the electron case. Electrons can spin either way, and the alpha particle has no spin and so it cannot. It is also necessary to make sure that the sum of all the probabilities add up to one, and so there has to be some normalization factor, N, that we have written out separately or which can be incorporated in the amplitude. In all the examples that we will discuss, $N\Phi(x)$ is constant or nearly so, and we will drop the x dependence and simply write Φ.

Exclusive alternatives can be written in a similar way as interfering alternatives. Equation (8) can be expressed in the same way as Equation (9) with $P_{L,R}(x) = \Phi|A_{L,R}(x)|$:

(3) The probability of outcome x for exclusive alternatives is

$$P(x) = \Phi_L|A_L(x)|^2 + \Phi_R|A_L(x)|^2 = P_L(x) + P_R(x). \qquad (10)$$

In both Equations (9) and (10), we multiply complex numbers by their conjugates. We pretty much have to! Probabilities must be real numbers between zero and one. There is pretty much only one way to convert a complex number into a nonnegative real number.

Notice that all this discussion is about probabilities. There is nothing about saying exactly what will happen, but only about saying what is likely to happen. The nondeterministic nature of quantum mechanics is probably the one single thing that bothers people the most about quantum mechanics.

For the situation in Figure 7.1, the amplitude is

$$A = e^{ips/\hbar} = \cos(ps/\hbar) + i\sin(ps/\hbar), \tag{11}$$

where s is the distance that the particle is traveling through space, and the number $\hbar = h/2\pi = 1.05457 \times 10^{-34}$ (kg m^2/s) is Planck's constant over 2π. It appears in every quantum mechanical motion problem. The momentum of the electron is p. Because $\hbar$ is so small, $p/\hbar$ is quite large. For simplicity, define $\kappa = p/\hbar$. A typical voltage in the electron gun might be 50,000 V; if that is the case, $\kappa = 1.17 \times 10^{12}$ m^{-1}. Because $p/\hbar$ is naturally big, small changes in s have a big impact. A change of 360° in κs for a 50,000 V electron happens in a distance s of only 5.4 picometers.

Already, we can start to see the mathematics of waves appearing in the forms of sine and cosine functions in Equation (11). That will come out more clearly when we compute the probability for the sum of the two amplitudes for the two-slit problem of Figure 7.1.

Equation (11) also illustrates a general property of amplitudes, which is that amplitudes for chronologically successive events multiply. If an electron travels a distance s and then travels a distance t, the combined amplitude is $e^{i\kappa(s+t)} = e^{i\kappa s}\, e^{i\kappa t}$.

Chapter 9

Your First Quantum Calculation

Now we work out the solution to the problem of the electron going through the two slits. Calculate first the amplitude $A_L(x)$ for the electron to go through the slit on the left and land at x. This distance x is not large; it will be comparable to δ in a typical experiment. The distance from the barrier to the screen is ℓ, which could be around 1 meter. The source is, let us say, distance d from the barrier. The total distance that the electron travels, s, is, using the Pythagorean theorem twice,

$$s = \sqrt{d^2 + \delta^2} + \sqrt{l^2 + (x - \delta)^2}. \tag{12}$$

The amplitude, as a function of x, to go through the left slit is

$$A_L(x) = e^{i\kappa(\sqrt{d^2+\delta^2}+\sqrt{l^2+(x-\delta)^2})}, \tag{13}$$

using $\kappa = (p/\hbar)$. The total amplitude, for the interfering alternatives of left and right slits is

$$A(x) = A_L(x) + A_R(x) = e^{i\kappa(\sqrt{d^2+\delta^2}+\sqrt{l^2+(x-\delta)^2})}$$
$$+ e^{i\kappa(\sqrt{d^2+\delta^2}+\sqrt{l^2+(x+\delta)^2})}. \tag{14}$$

According to the central procedure of quantum mechanics, we now take the amplitude written out in Equation (14) and multiply it by its conjugate in order to get the probability. Neglecting the factor of Φ, and limiting ourselves to $|x| \ll \ell$, the probability that the electron lands at point x on Figure 7.1 is given by $|A(x)|^2$, which is

$$P(x) \propto \cos^2(\kappa \delta\, x/l), \tag{15}$$

and which is plotted in Figure 9.1 for the case of $\delta = 1\,\mu\text{m}$, $\ell = 1\,\text{m}$, and an electron gun voltage of $50\,\text{kV}$.

Exercise 7. Get from Equation (14) to Equation (15).

Before pondering the meaning of Figure 9.1, notice that the distance d disappeared in going from Equation (14) to Equation (15). The distance between the electron gun and the slits is irrelevant to the location of the particle on the screen. This is a symmetry!

(1) *Say something describing what we observe*: In this case, the initial statement is to assert that Equation (15) will describe the probability of the electron appearing at x.
(2) *Transform what we are looking at in some particular way*: The transformation is to change the distance d.

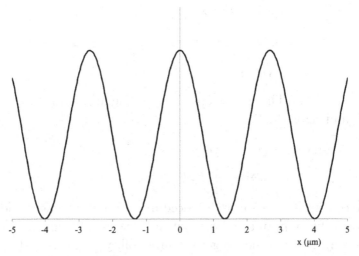

Figure 9.1: The interference pattern produced by the two slits of Figure 7.1, as determined by Equation (15).

(3) *Ask if we can still say the same thing we said in step 1. If so, we have symmetry*: And what we have worked out is that Equation (15) will still describe the probability of the electron appearing at x.

This symmetry is a special case of the principle of gauge invariance, which has turned out to be a very important symmetry. We will say more about it, and what it has to do with the Higgs later.

Chapter 10

Your First Quantum Experiment

The two-slit experiment has actually been done,[1] giving the results that we will describe here.

An electron leaves the electron gun at some point in time and it later appears at some place on the detector. It is going to land at some place where $P(x)$ is not zero; say it lands at $x = 3.4\,\mu$m for the particular values of δ and ℓ that we used to make Figure 9.1. Then another electron leaves the source and lands at some point, say at $x = -4.9\,\mu$m, and so on as more and more electrons make the trip. What we will see is little dots in our detector at different locations wherever the electrons arrive. When a large number of dots have appeared, they will be clustered at the places where $P(x)$ is large, and there will be only a few dots at places where $P(x)$ is small. Figure 10.1, from the paper by Merli, Missiroli and Pozzi shows what happens quite clearly.

The pattern of frame (f) in Figure 7.1 is the sort of pattern that arises frequently in the study of waves. This is a result of the specific amplitude of Equation (11), $e^{i\kappa s}$. Other amplitudes need not even have macroscopic distances such as s in them and need not have this type of explicitly wavelike nature.

Let us elaborate on the second example of interfering alternatives described earlier to make this a little clearer. If an electron and a

---

[1]P. G. Merli, G. F. Missiroli and G. Pozzi, "On the Statistical Aspect of Electron Interference Phenomena", *Am. J. Phys.*, 44, 306 (1976); A. Tonomura, J. Endo, T. Matsuda, T. Kawasaki and H. Ezawa, "Demonstration of Single-Electron Buildup of an Interference Pattern", *Am. J. Phys.*, 57, 117 (1989).

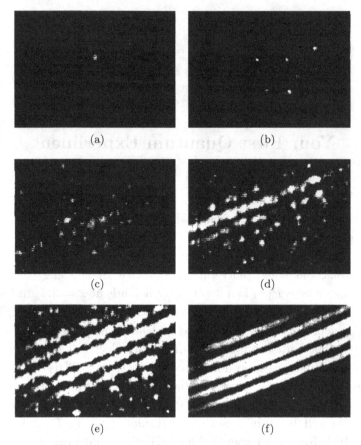

Figure 10.1: The appearance of electrons on a screen such as is shown on the top of in Figure 7.1; frames (a) through (f) are ordered with increasing time. Reproduced from *American Journal of Physics*, 44, 306 (1976), with the permission of the American Association of Physics Teachers.

positron annihilate to make a bundle of electromagnetic energy, and that bundle converts into a muon, μ^-, and an antimuon, μ^+, the process can be sketched out as shown on the left side of Figure 10.2.

In this Feynman diagram, the electron and its antimatter equivalent, the positron, e^+, collide to make a bundle of electric and magnetic fields. That is the squiggle line, a photon, γ. Pretty much instantaneously afterwards, the photon turns into a muon, μ^-, and an antimuon, μ^+. These go flying off in opposite directions.

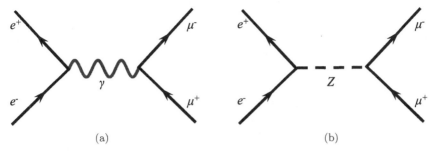

Figure 10.2: Two interfering alternatives for the annihilation of an electron and a positron to make a muon and an antimuon.

The cool thing about Feynman diagrams is that for each part of it, there are rules that make it possible to look at the diagram and write down what the corresponding amplitude is. Multiply the amplitudes for the parts, and that product will be the amplitude for the interaction represented by the whole diagram to happen. Let us start with the amplitude for the left side of Figure 10.2. It will have a factors of the form $e^{i\kappa s}$ for the incoming $e^{\pm}$; let us call those factors A_{IN}. There are similar factors for the outgoing $\mu^{\pm}$, which we will call A_{OUT}. In between is a factor for the bundle of electromagnetic energy and how it interacts with the incoming and outgoing particles. Call that factor A_{γ}. Because amplitudes for events which occur in succession multiply, the complete amplitude for the left hand side is then $A_{IN}\, A_{\gamma} A_{OUT}$.

The electron and positron can also annihilate to make a bundle of the fields involved in the weak force. That is depicted on the right side of Figure 10.2. The amplitude for this interaction is $A_{IN} A_Z A_{OUT}$, where A_Z is the factor in the amplitude to allow for the bundle of weak force energy and how that interacts with the incoming and outgoing particles.

Because it is not possible to tell which alternative occurred, these are interfering alternatives, and the probability for two electrons to become two muons is proportional to the magnitude of their sum, squared:

$$P(e^-e^+ \to \mu^-\mu^+) \propto |A_{IN} \propto (A_{\gamma} + A_Z)A_{OUT}|^2. \qquad (16)$$

Chapter 11

What Heisenberg Didn't Know

The central procedure of quantum mechanics says something about which alternative occurs, and in a way that we find bizzare, although not entirely impossible, to express in English. Normally, we think of the electron going through one slit or the other, just as we might enter a room or a house through one door or another. That is the case of exclusive alternatives; there is some probability for one thing to happen, and some probability for another thing to happen, and the probability for either to happen is just the sum of these two. Equation (8) expresses this neatly.

Equation (9) says some other thing. The probability is a sum of three things:

$$P \propto |A_L + A_R|^2 = |A_L|^2 + |A_R|^2 + 2\Re(A_L^* A_R). \qquad (17)$$

The first term, $|A_L|^2$ has only a number from the left hand slit in it, and the second term, $|A_R|^2$ has only a number from the right hand slit in it. These correspond to the exclusive alternatives of the electron going through either the left or right slit, individually. That really doesn't do much for us though, because the third term, $A_L^* A_R + A_R^* A_L = 2\Re(A_L^* A_R)$ has numbers from both slits involved in it. This term is the interference term, and it says in some sort of way that the electron went through both slits.

Now, what if we try to pin nature down on this? What if we try to set up the experiment so that we can tell which slit the electron went through? This can be done if there is some light going parallel to the sheet of metal with the slits in it. The light will bounce off the electron, and if you can measure that bounced light, you can, in principle, tell which slit the bounce happened at; that is the slit the electron went through.

Alas, this will create exclusive alternatives. In the case where the light says that the electron came through the left slit, the appearance of the flashes of light on the far screen will have a probability $P(x) \propto |A_L|^2$. When the electron came through the right slit, $P(x) \propto |A_R|^2$. There will not be any interference term. There also will not be any stripes on the screen as in Figure 10.1.

Exercise 8. Repeat the calculation of Exercise 7, but start with only one amplitude; $A(x) = A_L(x)$ say, without $A_R(x)$. You should get a result with no stripes or interference, i.e., that the probability $P(x)$ is a constant.

What happens is that the light bounces off the electron, but also the electron bounces off the light too. It no longer has the same energy and momentum as it did before it interacted with the light. As a result of the electron and light hitting each other, you know the position, but the electron's energy and momentum and direction are now different. You must start all over again to recalculate all the amplitudes with the electron beginning at the point where the light hit it. In this process, you will find that the interference has been destroyed, because there are not any indistinguishable alternatives to interfere between the point where the light hit the electron and the detection plane.

This reasoning has been tested experimentally. In 2007, a group in France[1] basically did this gedankenexperiment, except they used photons in place of electrons and an optical device called an electro-optical modulator in place of the detecting photons. They even arranged it so that the decision to make available interfering or exclusive alternatives took place after the photon entered the

---

[1]V. Jacques, E. Wu, F. Grosshans, F. Treussart, P. Grangier, A. Aspect, J.-F. Roch, "Experimental Realization of Wheeler's Delayed-Choice Gedanken Experiment", *Science*, 315, 966 (2007).

apparatus, or equivalently, after the electron left the gun. That highlights something quite bizarre about quantum mechanics, which we will come back to in The Quest for Meaning.

The principle here is that when a quantum mechanical system interacts with some other thing, the interfering alternatives prior to that interaction are destroyed. The interaction disturbs the system, and you must toss out the results of your previous amplitude calculations and recalculate a new set of probabilities according to the central procedure.

Werner Heisenberg realized this early in 1927; we will call this Heisenberg's Disturbance Principle. Because the interaction between the electron and the light is a measurement of where the electron is, the Disturbance Principle is fundamentally about the measurement process. When the electron is finally detected at the top of Figure 7.1, it exchanges energy and momentum with the screen; that, too, is an interaction or a measurement.

The Disturbance Principle very tightly limits our ability to know both the momentum and position of a specific electron with infinite accuracy. In lectures he gave in Chicago in 1929, Heisenberg said[2] "If the velocity of the electron is at first known, and the position then exactly measured, the position of the electron for times previous to the position measurement may be calculated". In that case, one knows the momentum for times before the measurement and the position at the time of the measurement. Perhaps one might argue that for the instant in which the interaction occurs, we know both, but it is far from clear how to validate that claim experimentally.

There is another, different, physical law which also limits our knowledge of the momentum and position of the electron. We will use the term Heisenberg's Uncertainty Principle for this different physical law, which does not involve the measurement process at all. The Uncertainty Principle is about a thing called the wavefunction.

So far we have worked with amplitudes for a system to go from one state to another. For example, we looked at the amplitudes by

---

[2]W. Heisenberg, *The Physical Principles of the Quantum Theory* (Chicago, IL: U. of Chicago Press, 1930).

which an electron and positron can become a muon and antimuon. There are also amplitudes for a particle in a single state to be "at" a certain location. An example of a corresponding classical quantity would be the probability for a pendulum to be at a certain point in its arc. Quantum mechanically, we have amplitudes rather than probabilities of course.

The wavefunction was originally just that: the amplitude for a particle to appear in a specific location. For the particularly important case of an electron in a hydrogen atom, there is some function with input $\vec{x}$, a position in space and output ψ, a complex number; $\psi(\vec{x})$ is a common way to write the wavefunction for that problem. As physicists began to understand that particles had other nonspatial properties such as spin, the idea of a wavefunction expanded to include these properties; and so, using σ for the spin (we need not be too precise here about defining σ), the wavefunction could be written as $\psi(\vec{x}, \sigma)$. The next thing that happens is that for some problem we are all excited about the amplitude for the spin to be a certain value and do not care about the position. Then we skip writing $\vec{x}$ everywhere and call $\psi(\sigma)$ the wavefunction.

Wavefunction is a bad name in the sense that when one is speaking of spin, or other nonspatial properties, the amplitude does not look like a wave. However, when one is speaking of position, $\psi(\vec{x})$ will have some resemblance to a wave. Wavefunction is a good name though in the sense that the amplitude is indeed a function; it takes some numbers such as position as input and produces a complex number as an output. The square of the magnitude of that complex number gives you the probability of finding the particle at the input position.

It turns out that a wavefunction that is a function of position, $\psi(\vec{x})$, contains more than information about just the probability to find the particle at position $\vec{x}$. Through a mathematical procedure call the Fourier transform, it is possible to determine the amplitude for that particle to appear with a specific momentum. The two amplitudes, $\psi(\vec{x})$ and $\hat{\psi}(\vec{p})$, are two different ways to say the same thing; if one knows $\psi(\vec{x})$, then one has determined $\hat{\psi}(\vec{p})$ and vice versa.

Heisenberg's Uncertainty Principle is a restriction on how sharply peaked the wavefunction can be for both the position and the momentum simultaneously. It is possible to construct a wavefunction that is as sharply peaked as you want in position. The magnitude squared of the amplitude can be quite high at some specific input position, corresponding to knowing the position as precisely as you want to know it. Alternatively, you could specify a wavefunction that has momentum precisely defined. Heisenberg's Uncertainty Principle says that you cannot construct a wavefunction that pins both down the position and the momentum (in the same direction, anyway) with perfect precision. There is not any combination of valid amplitudes which does that.

Here is an example. Figure 11.1 shows, on the top left, a particular wavefunction[3] for some particle. The wavefunction happens to be $\psi(x) \propto \cos(\kappa x) \exp(-x^2/2\sigma^2)$ for some particular values of κ and σ, but that is secondary. The corresponding probability distribution, which is not shown, is $\Delta x \cong 8.5\,\text{nm}$ wide. That is to say, this wavefunction localizes the particle to about 8.5 nm. We cannot tell from this though if the particle is moving to the left, the right, or indeed if it is moving at all!

Next, we apply the Fourier transform process and get $\hat{\psi}(p)$, the amplitude for this particle to have momentum p. That is shown on the top right side of Figure 11.1. We see that the probability for this particle to be going to the left is the same as the probability for it to be moving to the left, actually there is no amplitude for it to be stationary. More importantly, we see that a typical momentum, whether positive or negative, is $\Delta p \cong 34 \times 10^{-27}\,\text{kg m/s}$. Heisenberg's Uncertainty Principle says that the product $\Delta x \Delta p$ will be on the order of $\hbar$; it might be as low as $\hbar/2$ or it might be a bit more than $\hbar$, but by no means can it be zero. In this case, $\Delta x \Delta p$ is about 2.7 $\hbar$.

What happens if we try to localize the particle more? On the lower left we drawn a somewhat more localized spatial wavefunction, with $\Delta x \cong 2.1\,\text{nm}$ wide. We just changed the values of κ and σ,

---

[3]For all the cases shown in Figure 11.1, the wavefunctions and their Fourier transforms are real; there is no phase information.

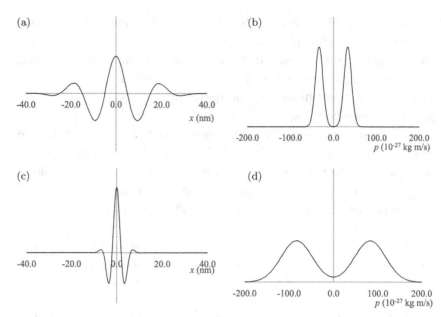

Figure 11.1: The wavefunctions described in the text. In the left panels the wavefunction is shown as a function of position; in the right panels, it is shown as a function of momentum. The panels on the top correspond to $\Delta x \cong 8.5$ nm and $\Delta p \cong 34 \times 10^{-27}$ kg m/s. The panels on the bottom correspond to $\Delta x \cong 2.1$ nm and $\Delta p \cong 86 \times 10^{-27}$ kg m/s.

actually. On the lower right, the same wavefunction, expressed in terms of momentum. Now the typical momentum is $\Delta p \cong 86 \times 10^{-27}$ kg m/s. This time, $\Delta x \Delta p$ is about $1.7\hbar$. When we change the spatial wave function so as to constrain the position better, the wavefunction in terms of momentum will spread out further, and vice-versa.

Despite their great similarity, the Disturbance Principle and the Uncertainty Principle really are two different things. The Disturbance Principle involves measurement; the Uncertainty Principle does not. And, because one can speak of the measurement of a single particle, it is simple to apply the Disturbance Principle to discussions of a single particle. The wavefunction determines only probabilities, and one must be very careful indeed when speaking of probabilities for single events.

Chapter 12

Gauge Invariance

The central procedure of quantum mechanics for the case of two interfering alternatives says that we compute probabilities from amplitudes using $P \propto |A_1 + A_2|^2$. There is an important symmetry hidden in the fact that probabilities are proportional to the magnitude squared of a complex number. This symmetry group is called $U(1)$, also known as the circle group. We met it back when we were discussing groups.

In Exercise 4, we described the elements of the circle group in terms on an angle of rotation ϕ, which was between 0 and 2π. For amplitudes, we use the form $e^{i\phi}$, with the operation for the group being the multiplication of complex numbers. In terms of our three formal steps:

(1) *Say something describing what we observe*: In this case, the initial statement is that for amplitudes A_1 and A_2, the probability is $P \propto |A_1 + A_2|^2$.

(2) *Transform what we are looking at in some particular way*: The transformation is to multiply both amplitudes by a number of the form $e^{i\theta}$. It also happens also that these numbers form a group.

(3) *Ask if we can still say the same thing we said in step 1. If so, we have symmetry*: What you have worked out is that after the multiplication, the probability P has not changed.

Exercise 9. In $P \propto |A_1 + A_2|^2$, multiply both amplitudes A_1 and A_2 by an element of the circle group. Show that the probability P does not change as a result of this transformation.

We have already seen an example of this in the two-slit problem; in going from Equation (14) to Equation (15), the distance d disappeared. The two individual amplitudes A_L and A_R both have factors of $e^{i\kappa\sqrt{d^2+\delta^2}}$ in them, as does their sum. This is the contribution to the amplitude for the motion of the electron from the electron gun to the screen with the two slits. When the magnitude squared, A^*A, is computed, these factors disappear; they do not appear in the expression for the probability. Physically what that is saying is that the interference pattern is entirely a result of what comes out of the two slits; the situation is the same as if we had two electrons sources in place of the two slits, and the formation of the sources really does not matter.

This symmetry, the symmetry of computed probabilities when the amplitudes are multiplied by a number of the form $e^{i\theta}$, is called gauge symmetry for historical reasons. It comes in many different forms, and the one that we have shown here is particularly, almost trivially, simple. Sometimes this simple case is called phase invariance. To be precise, this is a global gauge symmetry; we changed the phase of all the amplitudes everywhere by some certain specific constant amount. If the phase of all the amplitudes are changed by an amount that is a function of space and possibly time, that is a local gauge transformation. A local, rather than a global, gauge invariance is multiplication of the amplitude by $e^{i\theta(x,t)}$, where $\theta(x,t)$ is some function of space and time rather than some constant. The key thing here is that gauge symmetries are deeply embedded in quantum mechanics; the two are inextricably tied together.

It turns out that when writing the amplitude for situations where electromagnetism is involved, the phase of the amplitude depends on potential energies. Potential energies always have constants of integration, corresponding to arbitrary phase changes in the amplitude. One of the real breakthroughs in understanding the quantum nature of electromagnetic forces was the realization that the correct amplitudes for processes involving these forces have local gauge symmetry. That also was part of the story of how we came to understand the correct quantum formulation for the strong force. For the weak force, things were more complicated. That is why the solution of Peter Higgs and others (see A Historical Interlude) to incorporate gauge symmetry to the weak force was taken so seriously that we looked to find the Higgs particle for almost 50 years.

Chapter 13

Where Do the Quanta Come From?

The astute reader will notice that we have not actually said anything about quanta so far in our discussion of quantum mechanics. It is time to remedy that. In classical mechanics, an object can have any energy whatsoever. The energy of a planet as it moves around the sun in a stable orbit is given by a real number; it might be increased or decreased by any amount, even in principle an infinitesimally small amount. The energy of an electron as it moves around a proton in an atom in a stable state occurs only at specific, fixed values. In fact, all sorts of physical parameters that are considered as real numbers classically are expressed with integers when treated quantum mechanically. How does the central procedure of quantum mechanics create specific energy levels?

Sometimes, particularly when the position in space of a particle is part of the problem, quantization is the result of boundary conditions. A quantum dot will provide a good example. This is a device that will contain a single electron in a small box. Quantum dots can be manufactured in a number of ways; we will just imagine a one dimensional quantum dot with a span of π nm. Outside the range $-\pi/2\,\mathrm{nm} \leq x \leq +\pi/2\,\mathrm{nm}$, there are walls, and the wavefunction must be zero there.

It turns out — we shall spare you the mathematics — that the wavefunction is real and has no phase information; and in fact, it has to be sinusoidal. This, along with the fact that the wavefunction must be zero at the walls, limits what specific wavefunctions are possible. Figure 13.1 shows three possible wavefunctions for our quantum dot.

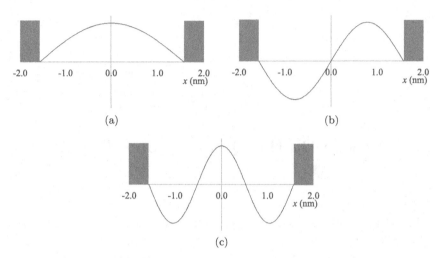

Figure 13.1: Wavefunctions for the three lowest energy levels of a one dimensional quantum dot of width π nm. The gray areas represent regions where the wavefunction is required to be zero because the electron is not allowed to be in the regions $|x| \geq \pi 2$.

Figure 13.1 shows three sinusoidal functions which are zero at the walls of the quantum dot. These are actually the wavefunctions of the three lowest possible energy levels. The energy levels happen to be $n^2(\hbar\pi/L)^2/2m$, where L is the dot's width (π nm in our case) and m is the mass of the electron. The number n is the quantum number for this case; it is an integer 1 or larger.

Exercise 10. Assume the de Broglie relation between the momentum p and the wavelength λ of the sinusoidal oscillations of Figure 13.1, $p = 2\pi\hbar/\lambda$. Using the classical expression $p^2/2m$ for the energy, derive $E_n = n^2(10^9\hbar)^2/2m$ for our quantum dot.

For the quantum dot, quantization happens because the wavefunction must go to zero at the boundaries of the dot. Similar requirements, similar boundary conditions, lie behind the quantization of many quantities. Boundary requirement quantize the energy levels of electrons in atoms and the vibrational energy of chemical bonds in molecules, for example.

There are cases where quantization occurs and we do not know of any boundary condition requirements. Electric charge is quantized in units of 1/3 the charge of the electron for example, but the boundary conditions that create that quantization are a mystery. It really makes physicists think that there must be some internal structure to electrons that has boundary conditions that force quantization of these sorts of quantities. For this reason, whenever a new accelerator facility makes it possible for physicists to search for an internal structure to an electron, they do so. So far though, none has been found, and the search has been conducted[1] down to distances of about 2×10^{-20} m.

---

[1]D. Bourilkov, "Hint for Axial-Vector Contact Interactions in the Data on $e^+e^- \rightarrow e^+e^-(\gamma)$ at Center-of-Mass Energies 192–208 GeV", *Phys. Rev. D*, 64, 071701(R) (2001).

Chapter 14

The Quest for Meaning:
Particles and Waves

Now, as fearless seekers of knowledge, we enter perilous waters.

In Chapter 11, we mentioned that the delayed-choice experiment of V. Jacques, E. Wu, F. Grosshans, F. Treussart, P. Grangier *et al.* highlighted something truly bizarre. In that experiment, the decision as to which paths were allowed — the decision as to whether interfering or exclusive alternatives were available to the particle moving through the apparatus — was made after the particle had already entered the apparatus. In terms of Figure 7.1, the electron somehow "knows" if there is a photon behind one of the slits that it will interact with, and adjusts the probabilities of different outcomes accordingly. However, the complete expression for the probabilities includes factors in the action for the motion of the electron before it reaches the slits. In terms of Equations (9) and (10), the correct mathematical expression to describe what will happen has to be determined in some way before the thing actually happens.

We present no explanation as to how nature actually does this. As far as we know, nobody does have a certain explanation, although people have been working at it.[1]

The bizarre nature of quantum mechanics has for a century caused inquiry about how we are to "understand" quantum mechanics. In some cases, these inquiries have gone well outside the range

---

[1]See, as an example, John G. Cramer, "The Transactional Interpretation of Quantum Mechanics", *Rev. Modern Phys.*, 58, 647 (1986).

of what could really be called science. In other cases, mathematical predictions about the motion of physical objects have been directly verified experimentally. The subject is not really closed; there is still a steady stream of serious research articles about some of the topics we will discuss here. There are still different schools of thought on the matter, and probably will be for some time. That is part of the reason few textbooks go into this subject in detail.

It must be admitted at the outset that this quest is optional. For all of the discussion and conflicting opinions about the interpretation of the wavefunction, it is quite possible to ignore the topic altogether. It is possible to use the central procedure of quantum mechanics, the mathematical devices of amplitudes and probabilities to compute predictions for the results of doable and repeatable experiments and see if your predictions come out correctly without really getting involved with these interpretation questions. This "shut up and calculate" approach, as it was called by David Mermin[2], is *by far* the most common approach among working physicists. That too is part of why few textbooks go into this subject in detail.

We will present the two most pivotal results in this field: the EPR paper of 1935 and Bell's Theorem from 1964. First however we will argue, though, that the wave-particle duality, at least as the idea was originally expressed, is an idea whose time has gone.

The wave-particle duality arose with, and to no small extent drove, the discoveries that created quantum mechanics. In de Broglie's 1929 Nobel lecture, he discusses how Planck, while studying black body radiation, had to assume that light was created and absorbed in equal and finite quantities, i.e., photons. Now at that time, the laws governing electric and magnetic fields were well known, and they clearly show that electromagnetic fields can form waves. The Bohr model of the atom and how it quantized electron motion was also on de Broglie's mind. Then, de Broglie writes,

> for both matter and radiations, light in particular, it is necessary to introduce the corpuscle concept and the wave concept at the same time. In other words the existence of corpuscles accompanied

---

[2] N. David Mermin, "What's Wrong with this Pillow?" *Phys. Today*, 42, 9 (1989).

by waves has to be assumed in all cases. However, since corpuscles and waves cannot be independent because, according to Bohr's expression, they constitute two complementary forces of reality, it must be possible to establish a certain parallelism between the motion of a corpuscle and the propagation of the associated wave.

The concept that the electron was simultaneously both a wave and a particle led de Broglie to the idea of "matter waves" and his famous relation $\lambda = h/p$, which in turn was a key ingredient in the subsequent development of quantum mechanics, including Schrödinger's equation and the subsequent development of chemistry as a result. It really is hard to overstate the impact of the duality idea has had on the development of quantum mechanics.

For this reason, a version of the wave-particle duality exists in nearly all introductory textbooks, although it is not always stated in the particular form that de Broglie advanced.

De Broglie was thinking something that is not in the two-slit calculation we did earlier. In our calculation, we determined the probabilities of an outcome based on amplitudes of the form $\cos(\kappa x) + i \sin(\kappa x)$. In deBroglie's discussion, the electron actually *is* a particle and it actually *is* an associated wave. Admittedly, in determining the probabilities of the electron to appear at a certain place, we are implicitly saying something about an electron's motion, and by starting from $e^{i\kappa x}$ in the amplitude we are implicitly saying that the motion is wavelike. To say that the electron's motion is wavelike is however a weaker statement than the claim that the electron actually is a wave; at most it says only that there are wave effects in its propagation.

Are we even really talking about a scientifically resolvable issue? Consider the electron in flight between the slits and the detector of Figure 7.1. It is a point sized entity when it interacts with the detector, as seen in Figure 10.1. What is it just an infinitesimal instant before the moment it interacts with the detector? Can we say that it is also a wave, and that somehow this wave disappears at the time of measurement, at the time when the electron interacts with the detector? Alternatively, can we say it is not also a wave? It does go unobserved in this time.

One might try, in a simple minded way, to attempt to observe the electron all the time. One could flood the volume of space through which the electron is moving with light and track from the reflected light where the electron is. The number of points on the track can increase without limit as the intensity of the light is increased. From the Disturbance Principle, the momentum, energy, and direction of the trajectory of the electron will be changed at each interaction, so the track will be far from a straight line, but this can indeed be done.

In fact, something similar to this happens when a high energy electron is sent through matter. Instead of the electron being flooded with light, it is flooded with the electric fields of the electrons in the molecules of the gas. The high energy electron bounces off these fields, and the fields bounce off the high energy electron, so to speak; the high energy electron interacts with the electric fields from the molecular electrons, and the molecules of the gas are thereby ionized. Measuring the trail of the ionization created by a charged particle as it passes through matter is the essence of every method of detecting and measuring particles.

The idea of flooding is suggestive, but not conclusive. No matter how short the intervals of time between interactions are, those intervals still exist. In those (short) intervals we could, if we chose, still consider the electron to be both particle and wave.

A stronger argument against the idea that the wave is physically real may be found in our second example of interfering alternatives, the two processes of Figure 10.2. The electron, positron, muon and antimuon will have wave effects as they propagate towards or away from each other, but the process whereby the electron and positron become a muon and antimuon has no wavelength, no frequency, and no interference patterns in the traditional sense — the particles do not travel from one place to another. Wave effects are not universal in quantum calculations.

We would like to like present another reason why regarding the wave effects of quantum motion as resulting from the existence of a real, physical wave associated with the point like electron is erroneous. First however, a brief diversion.

Chapter 15

The Logos

Uncle Leo came over to visit Melissa, who was needing some encouragement with her middle school mathematics. He put three pencils on the table. "Where is the three? Is it under this one?" he asked, lifting one of the pencils. "Is it inside this one?" he asked, disassembling the mechanical pencil and dumping the graphite leads out all over the table.

"Uncle Leo", Melissa replied, "Whaaaat?"

"The three is in your head. It is an abstract thing. This thing that you draw," (Uncle Leo picked up the remaining pencil and drew 3) "that is a word, a name for the thing, just like this," (Uncle Leo drew three) "or this," ($\equiv$, which is about the extent of Uncle Leo's ability to write in Chinese). "So if you write $3 = \equiv$, then what you are saying is that the thing that has this name 3 is the same as the thing that has that name $\equiv$, even though the thing does not really exist anywhere except between your ears!" Uncle Leo began to grow enthusiastic about what was obviously one of his favorite topics. "And this is the amazing thing! I have the same thing also between my ears, even though that thing, that 3, does not exist anywhere in the world! But we both have it in our heads! The same thing! Is that not totally astonishing?"

Melissa looked at Uncle Leo and blinked. Several times.

Since the time of the ancient Greek philosophers at least, people have wondered why words are an effective way to reason about and understand our world. Perhaps you could claim that if human brains did not make good mental models of the world around us, then humans would not have survived the evolutionary process. That sort of makes sense but it is not quantitative, and it seems hard to validate experimentally. Alternatively, there is the approach of Heraclitus and the Apostle John. That approach is not science either, but it is not pretending to be. But that is a topic for another book, which probably has the word Metaphysics, not the word Physics, in the title.

The particular type of words that are mathematical symbols seem to be very effective in describing the universe. Mathematics is, as Eugene Wigner phrased it[1], unreasonably effective in describing the natural world. Galileo went further[2], asserting that mathematics is the specific language in which the laws of physics are written.

> Philosophy is written in this grand book, which stands continually open before our eyes (I say the Universe), but we cannot understand without first learning to comprehend the language and know the characters in which it is written. It is written in mathematical language, and its characters are triangles, circles and other geometric figures, without which it is impossible to understand a word; without these one is wandering in a dark labyrinth.

We have already seen something of this in the two-slit experiment, with the interference term of Equation (17), and how it is able to easily express in a very precise way something that just does not fit well into more ordinary languages.

Here is another example of the curious power of mathematics to describe the universe. At the time of the French Revolution, Evariste Galois was studying the properties of polynomials and developed the first notions of groups. Galois knew that many physical phenomena were explained by polynomials, but he really was just wondering which kinds of polynomial equations had which kinds of solutions. Nevertheless, as we are showing you, groups became a central part of all sorts of physics in the 20th century, with uses ranging from the structures of crystals to the properties of subatomic particles.

---

[1]E. Wigner, "The Unreasonable Effectiveness of Mathematics in the Natural Sciences", The Richard Courant Lecture in Mathematical Sciences, delivered at New York University, *Commun. Pure Appl. Math.*, 13, 1 (1960).
[2]G. Galilei, *The Assayer* (Rome: Giacomo Mascardi, 1623).

So a kind of mathematics, developed with no particular thought of describing the universe turned out much later to describe it well, and to describe many different aspects of it. Moreover, it describes facets of the world that are unexpectedly different from the context in which the mathematics was created.

Chapter 16

Mental Waves

Amplitudes are clearly mathematical rather than physical objects. Like the number 3, they are intellectual constructions that describe reality, and not the reality itself. For one thing, they are manipulated mathematically to make probabilities, and the mathematical nature of a probability is indisputable.

A key difference, perhaps the defining difference, between mathematical objects and concrete ones is that concrete objects travel at the speed of light or less. Mathematical things are not limited in that way.

Imagine a pair of rather large pair of scissors, as shown in Figure 16.1. Imagine a pair of scissors with blades on the order of 10^{21} m long — about the size of the Milky Way, for example. The blade tips move vertically. The intersection point between the two blades moves horizontally. The vertical motion of the atoms in the scissor blades, the atoms of iron and carbon, is constrained to be less than the speed of light. It is not possible to move a massive, concrete object faster than that; the energy required to do so is infinite. The horizontal motion of the intersection point of two blades however is not so constrained. Indeed, the intersection point moves with a speed approaching infinity as the scissors close. The intersection point is a mathematical object; it is the intersection of two lines.

At the very instant when the electron is detected at some point x at the top of Figure 7.1, the probability for it to appear at any other point x' goes immediately to zero. This is true even if that point x' was so far away from x that faster-than-light travel would be needed

Figure 16.1: A rather large pair of scissors, on the order of 10^{21} meters in length. Courtesy Zoë Steele.

to get from x to x'. At the moment when the electron is detected, all our calculations about the probabilities for it to appear in any other place instantly become irrelevant.[1]

Maybe another example here will help. Our senior scientist is having a senior moment, knocking around in the closet looking for his hat. "Fifty-fifty chance I left it back at the hotel," he thinks. The hotel is in a city 3000 kilometers away. Light takes 10 milliseconds to travel that distance. At the stroke of noon, he turns around to find his beloved holding the hat, saying "Looking for this?" The 50% probability that his hat is at the hotel goes to 0% probability. It goes to 0% not in 10 ms, but in no time at all. On the other hand, had his beloved not been standing there holding the hat (and smiling, oh yes) then our senior scientist would have phoned the hotel and asked for the hat. Then a signal, a concrete bundle of electromagnetic energy,

---

[1]We are being careful here in our choice of words. Mental processes in humans happen over a timescale of hundreds of milliseconds at best and order $\gtrsim 10^8$ seconds in some cases. We are not saying anything about how long it takes for any of us to realize that the result of the existing probability calculation is irrelevant, only that it in fact is irrelevant.

would have had to go to the hotel and back over some wires of some kind of radio signal. That trip takes 20 ms at least.

Now there is probably someone reading this who loves probability theory and is in an uproar. We are using here the Bayesian concept of probability. Bayesian probability is a quantitative expression of a person's belief about the nature of reality. A fair number of physicists do not like such a thing. Science is supposed to be about reality, not about somebody's opinions about reality. The other approach to probability is the frequentist definition. In the frequentist approach, the experiment is repeated (or imagined to have been repeated) a large number of times and the probability of some outcome is the fraction of those times in which that outcome happens. In the two slit problem, we are talking about a sample of one, rather than many, electrons passing through the apparatus, and the Bayesian language is simpler.

Interestingly, the whole process of computing a probability in quantum mechanics does not depend on which definition of probability you use. Historically though, nearly all the developers of quantum mechanics have worked with the frequentist interpretation. It is also worth noting that both Bayesian and frequentist probabilities are both clearly mathematical rather than physical objects; and they are the result of applying a mathematical operation to the amplitudes.

If deBroglie waves are concrete, physical things, then the appearance of the electron at x must change the shape of a concrete, physical wave. That changed shape has to propagate from x out to all other places. Because the electron did appear at x, it cannot appear at any other place x' and so the deBroglie wave must go to zero everywhere but at x. A physical signal, so to speak, has to propagate out from x, where the electron actually appears, to all other places x' to cause the deBroglie wave to collapse.

However, that propagation takes time. When the electron appears at x, that is analogous to our scientist finding his hat at home. An electron appearing at x' is analogous to our scientist discovering that he did indeed leave his hat at the hotel. A physical signal, such as the collapse of a physical deBroglie wave, takes 10 ms or more to proceed from the house to the hotel. After 1 ms, the wave has not gone to zero at the hotel and it is therefore still possible for the hat to appear there. Now we have created two hats, or two electrons.

Surely that cannot happen! And indeed, it does not happen. While this specific experiment has not been done with electrons, it has been done with photons,[2] with the result that photons (and no doubt, hats) are conserved. A physical, spatially extended wave of the sort that deBroglie was thinking of has to collapse instantaneously upon measurement at x; but a physical, spatially extended object cannot collapse instantaneously. Indeed, it is erroneous to think of a mathematical object such as the wavefunction's value at x', or indeed any function $f(x')$ as being physically at the point x' in the first place.

---

[2]T. Guerreiro, B. Sanguinetti, H. Zbinden and A. Suarez, "Single-Photon Space-Like Antibunching", *Phys. Lett.*, A376, 2174 (2012).

Chapter 17

Einstein, Podolsky and Rosen

Quantum mechanics differs from classical mechanics in quite a few ways. First and foremost, the idea of two objects or events being perfectly indistinguishable is central to quantum mechanics and does not really play much role in classical mechanics. Closely tied to this is the idea that certain questions simply can not be answered quantum mechanically: we cannot know which of two interfering alternatives occurred, we cannot know the exact position of a particle without measurement, and if we do measure it, we disturb its trajectory and do not know where it will be next. If we have two particles in close proximity, we cannot say which one is which, because they are perfectly identical. But of all the things that we cannot know in quantum mechanics, the one that probably bothers people the most is that we cannot know what is going to happen next.

Physical motion on the human scale is, at least in principle, predictable. Throw a baseball at a certain precisely specified speed with a certain specific spin and hit it at a certain specific angle with a bat of some specific mass and speed, and the outcoming trajectory of the ball can be computed. This repeatability is why all sorts of machines, from automobiles to assembly lines, can be made to work reliably. But for quantum motion, no such thing is true. We can calculate the probabilities of different possible outcomes, but we cannot say which of the different possible outcomes will actually occur. What is it that causes one electron to appear here on the screen and the next electron, coming out of the same electron gun

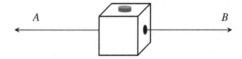

Figure 17.1: The Einstein, Podolsky and Rosen machine.

and going through the same slits, to appear at some other place on the screen? It really feels like quantum mechanics is not a complete description as to what is happening, because it has no answer for this question.

The landmark attempt to prove that quantum mechanics could not be complete was a paper by Einstein, Podolsky and Rosen[1] in 1935, which is often called just "EPR".

EPR supposed that we have a device of the sort shown in Figure 17.1. When the button on the top of the box is pressed, a particle at rest is produced inside of it, at some specific location. That particle decays into two particles that head off in opposite directions. We wait until the two particles, called A and B, are so far apart that it is sensible to speak of the wavefunction of A and the wavefunction of B as two separate things. This must happen; at some point, when A is very far from B, we would have no problem to write and speak of the wavefunction of A with no reference at all to how it was created. We would be able to measure the position or, should we choose, the momentum, of A, and the result of that measurement should be determined by that wavefunction.

Correctly speaking, a single measurement of this type will not determine anything about the wavefunction. What we have to do is repeat the measurement a large number of times, record all the measurements, work out the probabilities, and that result should be in agreement with the wavefunction of A.

However, A and B are correlated. If we measure the momentum of A to be some value, we know that B must have a momentum of

---

[1]A. Einstein, B. Podolsky and N. Rosen, "Can Quantum-Mechanical Description of Physical Reality be Considered Complete?", *Phys. Rev.*, 46, 777 (1935).

the same magnitude[2] in the opposite direction. On the one hand, since A and B are far apart when they are measured, it is sensible to think of them as separate objects. On the other hand, since you learn something about B by looking at A, the two wavefunctions are not completely independent, either. Occasionally people will say that these are "entangled", which may be true but is not the essential point. The essential point is that if you learn something about B by looking at A, they are correlated. The word "entangled" refers to a specific way of creating correlation that is intrinsically quantum mechanical. However, correlations that are not quantum mechanical are quite possible.

Consider Reinhold Bertlmann's socks.[3] As it happens, Reinhold never wears socks that match, and which color he will have on a given foot on a given day is quite unpredictable. Accordingly, if you observe that one of Bertlmann's socks is green, then you are immediately certain that the other sock is not green. In this case, both the momenta and the positions of A and B are correlated. In both the case of Bertlmann's socks and the EPR apparatus, the correlations are a result of how the two subsystems are created.

There is a caveat regarding the knowledge we get from correlations in this common-origin situation. When we measure particle A, when we observe Bertlmann's left sock, we "know" something about particle B, about Bertlmann's right sock, but only in a limited sense. There is no physical carrier of information, moving from the measurement of particle A to that of particle B and affecting the measurement of B. Remember that physics is an experimental science, and the most fundamental law of all experimental sciences is

---

[2]Curiously, EPR did not discuss the effect of the Uncertainty Principle on the initial state. This point was immediately raised by N. Bohr, "Can Quantum-Mechanical Description of Physical Reality be Considered Complete?", *Phys. Rev.*, 48, 696 (1935). The replacement of position and momentum with spin coordinates, makes the point moot. This is because the spin of the initial state, unlike the position or momentum, is quantized discretely and can be set to zero, exactly.

[3]J. Bell, "Bertlmann's Socks and the Nature of Reality", *J. de Physique Colloques*, 42(C2), 41 (1981).

Murphy's: anything that can go wrong, will. At the moment when we measure particle A, something can go wrong, and the measurement of particle B can fail. In fact, measurement at A does not let us know that B even exists anymore, never mind that it has some particular measurable property. Equivalently, knowing that Bertlmann's left sock is green does not rule out the possibility that somewhere in the course of the day, Bertlmann has lost his right sock, perhaps on some foolish wager.

We can however circumvent this caveat by repeating the experiment many times, discarding repetitions where the apparatus failed, and studying the remaining data.

Now we come to the heart of EPR's argument. Suppose we repeat the experiment many times — say 1000 times. Half of the time, we measure the position at A; for the other 500 repeats, we measure the momentum at B. We write down the results, throw out the data where something went wrong with the apparatus, and contemplate the meaning of the results.

All 1000 runs are identical. The wavefunction of A does not change, nor does the wavefunction of B. We can use the 500 position measurements to infer the spatial distribution of the wavefunction of A, and the 500 momentum measurements to infer the momentum distribution of the wavefunction of A, and of course those two distributions must obey the Uncertainty Principle.

So far so good. Now we move on to a second experiment. We operate the EPR machine again, but only enough times to get one good piece of data. If the apparatus is reliable, we might get good data on the first push of the button! We measure the position at A, but we also measure the momentum at B. We do not specify the time when B is measured; maybe the momentum measuring device is close to the box, or maybe it is far away. Now, when we measured the position at A, because of the correlated way in which A and B were produced, we know, although we did not measure, the position at B exactly. And we measured the momentum at B, and so now we have both the position and momentum of B, and we have them with as much precision as we want.

Because there is a measurement at B, we are not asking if the Uncertainty Principle is violated; that is not the issue. The Uncertainty Principle is about the unmeasured wavefunction. Furthermore, the Disturbance Principle is not violated. The motion of B after it is measured cannot be predicted perfectly; the motion of B is indeed disturbed by the measurement. The claim of EPR is only that we know both the position and momentum of B at some instant of time, the time when B was measured. Since both quantities are known, then they must both, in some sense of the word, be considered physically real; they must exist.

EPR did not benefit from the hindsight that permits us to distinguish the Disturbance Principle from the Uncertainty Principle. Instead, they were working with the idea, which probably originated around 1925 during Heisenberg's stay in Copenhagen with Niels Bohr, that one must not speak of or incorporate into the theory things that cannot be observed; and as momentum and position cannot be observed at the same time, the interpretation was that they could not exist simultaneously. As EPR showed that they did, in a certain situation, exist simultaneously, EPR concluded that quantum mechanics was incomplete. It must be, they argued, that position and momentum are real entities, and the fact that quantum mechanics could not figure out what they were was quantum mechanics' fault.

Suppose then that we accept the argument of EPR, that quantum mechanics cannot be a complete description of reality. In that case, there must be variables of some sort, hidden to us, that in some way define the momentum and position of each of the two particles. The simplest hidden variables would be the position and momentum themselves, but that brings us back to classical mechanics again. Some other variables that would determine position and momentum, without actually being them, are the sort of things we are looking for. The true and correct theory then would be described with these hidden variables and, Einstein hoped, be fundamentally deterministic. The quantum theory would simply be an approximation to this true theory, and the probabilistic nature of its predictions would not be a fundamental property of nature. If there are hidden variables, then we would be saved from the thought that, as Einstein put it, God plays dice with the universe.

In 1964, John Bell set forth an experimentally verifiable statement that showed that within quantum mechanics, there are situations where hidden variables are impossible. And the experiments have indeed been done, and hidden variables are impossible. Before getting into the details of Bell's work, we need to know a little more about spin, and we need a simple logical theorem.

Chapter 18

About Spin

We first discussed spin in A Symmetry That Is Not. The study of spin is a fascinating topic in its own right; one could write a small book on the topic. What we need to know for Bell's Theorem is about the spin of electrons.

Revolving objects have angular momentum. If the object is heavier, or spinning faster, it will have more angular momentum than if it is lighter or spinning more slowly; that is expressed in the magnitude of the angular momentum. Angular momentum is a conserved quantity; an object that is rotating around some axis will continue to rotate around that axis unless acted upon by some outside force applied to the object at some point off the axis of rotation.

There is a direction to angular momentum, as well. The direction of the axis of the rotation also matters. The front wheel of a bicycle, rotating around its horizontal axle, is different from that same wheel being rotated around a vertical axis by its handlebars. The axis of rotation points in a specific direction and so you might think that angular momentum, having both magnitude and direction, is a vector. Before we quantize, that is fine. After we quantize, we will not be able to say that.

There is a minor complication in that an axis of rotation is a line and as such it points in two opposite directions, not one. The convention is to say that the direction of the vector is along the axis of rotation and is defined with the righthand rule. Curl the fingers of your right hand in the direction of the rotation, and your thumb will point in the direction of the angular momentum. The direction of the angular momentum of the hands of the clock is into the wall. In Figure 3.1, the spin of the $\bar{\nu}$ is in the same direction as its momentum, that is, toward the top of the page. The spin of the electron is also toward the top of the page.

Molecules, atoms, nuclei and subatomic particles can have angular momentum for two reasons. The first is just that they are rotating around their center of mass; a CO_2 molecule, which has the carbon atom sandwiched between the two oxygen atoms, might be spinning around the carbon center. The second is that subatomic particles can also have some angular momentum intrinsic to themselves. In the case of elementary particles like electrons, we do not really understand why this is true, any more than we understand why the charge on an electron is what it happens to be or why the mass of an electron is what it happens to be. This kind of angular momentum is called the spin of the particle.

The units of angular momentum are (kg m^2/s) — just the same as Planck's constant, $\hbar$.

On the scale of molecules, atoms and particles, spin and angular momentum are quantized. They can change only in units of amount $\hbar$. In the case of a rotating object, such as a CO_2 molecule, the angular momentum must be an integer multiple of $\hbar$. For spin, this happens, but there is another possibility as well. Some particles have spin $n\hbar/2$, where n is an even number; that is the same as an integer n times $\hbar$. These particles are the bosons. There are other particles with spin $n\hbar/2$, where n is an odd number;[1] these are the fermions. We will say more on bosons and fermions later, in Two Fermions in a Pod.

When dealing with angular momentum at the quantum level, the vector idea is not quite right. All three components of a vector can be defined precisely with three real numbers at the same time. However, at the quantum level, it is not possible to specify all the components of angular momentum at the same time. The situation is akin to, but not exactly the same as, the Uncertainty Principle.

---

[1]To be precise, these are the components of the spin in some particular direction, rather than the total magnitude of the angular momentum.

Although we do not have a standard conventional vector in the quantum world, there is a thing that we can draw as an arrow that will describe the spin of an electron. It will look like a vector when we draw it, but it will not have three independent components. It will have only an angle relative to a specific direction. We will draw it as an arrow, but use a dotted line, and call it an almost-vector rather than a vector.[2] Its length will be $\hbar/2$; the electron is a fermion, with spin $n\hbar/2$, where n is an odd number, to wit, 1.

Figure 18.1 has a classical vector $\vec{L}$, and its component in the arbitrarily selected x direction, which is $L_x = |\vec{L}|\cos\theta$. Figure 18.2 is the quantum version, for the electron. In this case, the component of the spin in the x direction is always measured to be either $+\hbar/2$ or $-\hbar/2$. Which one is actually measured is matter of probability. The probabilities of getting the different values are related to the angle θ. For $\theta = 0$, the probability of getting $+\hbar/2$ is 1 and of getting $-\hbar/2$ of course is 0. For $\theta = \pi$, the probability of getting $+\hbar/2$ is 0 and of getting $-\hbar/2$ is 1. For $\theta = \pi/2$, the probabilities of the two outcomes are equal. The general rule is that the probabilities will be so that the average of many equivalent readings is the value you get by projecting a classical vector of length $\hbar/2$ at angle θ to the

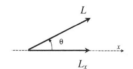

Figure 18.1: The projection of an angular momentum vector.

Figure 18.2: The projection of a spin $\hbar/2$ almost-vector.

---

[2]Warning! Many texts draw things that are like vectors for quantum spin which are a bit different from our almost-vectors.

x axis. That is, the expected value of projecting the almost-vector is $L_x = |\vec{L}| \cos \theta$.

Suppose we have a supply of electrons, all with the same spin direction. If we pick the direction x so that the spin is always measured as $\hbar/2$, then our almost-vector is at angle $\theta = 0$ relative to the x direction. Pick also a second direction, y, so that the almost-vector is at some nontrivial angle θ relative to the new axis. Then, to get the average reading of the spin in this new direction to equal what we would get by projecting a vector of length $\hbar/2$ through the angle θ, we must have that

$$\left(\frac{+\hbar}{2} \right) P_{+\hbar/2} + \left(\frac{-\hbar}{2} \right) P_{-\hbar/2} = \left(\frac{\hbar}{2} \right) \cos \theta. \tag{18}$$

where $P_{+\hbar/2}$ is the probability of measuring a spin of $+\hbar/2$ and similarly for $P_{-\hbar/2}$. Substituting $P_{-\hbar/2} = 1 - P_{+\hbar/2}$, and using fact that the probability is the square of the amplitude,

$$P_{+\hbar/2} = |A_{+\hbar/2}|^2 = \frac{\cos \theta + 1}{2}. \tag{19}$$

where $A_{+\hbar/2}$ is the amplitude for the electron to appear with spin in the x direction of amplitude $+\hbar/2$.

Exercise 11. Show that, neglecting a possible phase, the amplitude for an electron with the almost-vector of Figure 18.2 to be measured with a spin in the x direction of $+\hbar/2$ is $\cos(\theta/2)$, and that the amplitude for a measurement of $-\hbar/2$ is $\sin(\theta/2)$.

Chapter 19

Bell's Theorem: Setting up the Equipment

Consider a set of objects each of which has, or does not have, three attributes. If we use the example of Bertlmann's socks, we might have that any given sock either is green (G) or is not green $(\neg G)$; and either it is striped (S) or is not striped $(\neg S)$ and either it has holes (H) or does not have holes $(\neg H)$. Then, pulling socks out of the drawer at random, there are certain probabilities for each combination. For example, the probability of a green sock without holes is $P(G, \neg H)$ and the probability of a nongreen striped sock with holes is $P(\neg G, S, H)$. The theorem we will need is

$$P(G, \neg S) + P(S, \neg H) \geq P(G, \neg H) \tag{20}$$

Exercise 12. Prove Equation (20).

Equation (20) is Bell's Theorem or Bell's Inequality, although it is not exactly in the exact same form as in Bell's paper.[1] The key assumption to Equation (20) is that we can meaningfully talk about a sock being either green or not green, striped or not striped and having or not having a hole. What we are going to do is show a specific quantum mechanical situation where Equation (20) is *not* true, and then conclude that this key assumption can not therefore be made.

Figure 19.1 shows depicts the gedankenexperiment that we will use as a test of Bell's Theorem. Bell uses an EPR machine,

---

[1] J. Bell, "On the Einstein Podolsky Rosen Paradox", *Physics*, 1, 195 (1964).

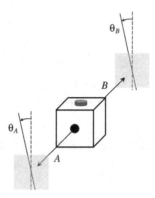

Figure 19.1: Bell's Theorem gedankenexperiment. The angles of the spin measurements relative to the vertical are θ_A and θ_B.

but specifically, he uses an EPR machine that emits electrons. At some distance from the box, we measure the components of the electrons' spins in the directions given by θ_A and θ_B.

The spins of the two electrons are created by the box to be in opposite directions. They are correlated, like Bertlmann's socks; you can learn about the spin of one electron by observing the other electron. But they are more than just correlated; they are entangled.

The electrons are created in a specific quantum state, called the singlet state. The total spin of the singlet state is zero in any direction. If $\theta_A = \theta_B$, events where a spin $+\hbar/2$ is measured at A are events where the spin must, with 100% probability, be measured as $-\hbar/2$ at B. But, because a measurement of the spin must give either $+\hbar/2$ or $-\hbar/2$ and no other values, the singlet state has some unusual properties when $\theta_A \neq \theta_B$, as we will see. Singlet states are quite common; the two electrons in an He atom will form a spin singlet. Singlet states are part of a larger category of states which are called entangled. Entangled states are, in a sense, the quantum mechanical version of Bertlmann's socks.

Events where $+\hbar/2$ is measured at A are events where we know that the almost-vector for electron B must be pointed in the direction $\pi - \theta_A$. The measurement of the spin at B is controlled by the amplitudes we worked out in the answer to Exercise 11. The result $-\hbar/2$ happens with amplitude $\cos((\theta_B - \theta_A)/2)$, because the

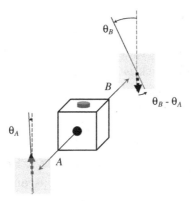

Figure 19.2: Bell's Theorem gedankenexperiment, after an observation of a spin of $+\hbar/2$ at angle θ_A on side A. The thick dotted arrows are the almost-vectors of the two electrons, as determined by the measurement at A. The almost-vector at B is at an angle $\theta_B - \theta_A$ relative to the direction in which it will be measured.

angle between the almost-vector and the direction of measurement is $\theta_B - \theta_A$. See Figure 19.2. The probability for a "backwards" or "same sign" reading, where both A and B read $+\hbar/2$ (or equivalently, for both to read $-\hbar/2$) is the square of $\sin((\theta_B - \theta_A)/2) \cong (\theta_B - \theta_A)^2/4$.

Consider three hidden variables — analogs to our sock variables of being green, being stripes and having holes. These hidden spin variables are defined by moving the angle of measurement of the electrons' spins away from vertical. In real life, Bertlmann's socks were correlated in color only. For our test of Bell's theorem in the quantum world, we postulate that they are also correlated in stripes and holes.

The first hidden variable is defined as true when the spin of the A electron is measured to be $+\hbar/2$ if the direction of measurement is straight up ($\theta_A = 0$). Equivalently, the first hidden variable is true if spin of the B electron is measured to be $-\hbar/2$ when the direction of measurement at B is also straight up ($\theta_B = 0$). This case is analogous to observing that the left sock is green or equivalently, that the right sock is observed to be not green.

The second hidden variable is defined as true when the spin of the A electron measured to be $+\hbar/2$ when the direction of measurement points is at some small angle s to the vertical ($\theta_A = s$). Equivalently, second hidden variable is true if the spin of the B electron is measured as $-\hbar/2$ when the direction of measurement at B is $\theta_B = s$. This case is analogous to observing that the left sock is striped or equivalently, to observing the right sock is not striped.

The third hidden variable is true when the spin of the A electron on the left is measured to be $+\hbar/2$ and the direction of measurement points off at some less-small angle θ_A (call it t). "Less small" specifically means that $t > s$, but t is still small enough that we could use the small angle formula $\sin(t) \cong t$. Remember that $t > s$. That will be important later. This case is analogous to observing that the left sock as having a hole or equivalently, to observing the right sock as not having a hole.

Notice that we did *not* speak of the spins, or the socks actually *having* properties in an intrinsic way. We did not speak of the socks as being green or having holes, or of the electron having or not having one of the three properties when we are not looking at them. We only spoke of what was measured at A and B. We are defining measurable quantities that can be interpreted as hidden variables without just making them hidden variables at the outset.

Chapter 20

Bell's Theorem: Taking the Data

Now we can work out a quantum form of Bell's Theorem in the form of Equation (20). Or try to! We take three sets of data with our experiment. Each set consists of repeatedly pressing the button on the EPR machine, measuring the components of the spins at A and B, and recording those measured spins. Using the frequentist definition of probability, we will process the recorded data to determine the probability of each particular combination of measurements occurring.

First, take data when $\theta_A = 0$ and θ_B is some small angle s. That is, we check on side A for the first hidden variable, and on side B for the second hidden variable. This is analogous to checking on the left side for a green sock and on the right side for stripes. When A reads $+\hbar/2$, you will see B read $-\hbar/2$ most of the time, because s is a small angle. But because s is not zero, there will also be a few same-sign readings at B, when B gives $+\hbar/2$. That will happen with probability $(\theta_B - \theta_A)^2/4 = s^2/4$.

The probability that A reads $+\hbar/2$ is $P(G)$, the probability of the left sock being green. If B reads $-\hbar/2$ when $\theta_B = s$, then B has no stripes, which implies that A has stripes. The probability that B obtains a same-sign reading of $+\hbar/2$ is the probability that B indeed has stripes, which is the probability that A has no stripes. In short, $P(G, \neg S)$, the probability that the left sock (the electron at A) is green and unstriped (has the first but not the second hidden variable) is $s^2/4$.

The second and third data sets are the same as the first, but the angles are selected differently.

The second data set is taken when $\theta_A = s$ and $\theta_B = t$. This is analogous to checking on the left side for stripes and on the right side for holes. The conclusion is that $P(S, \neg H)$, the probability that the left sock (the electron at A) has stripes but has no holes (has the second but not the third hidden variable) is $(t - s)^2/4$.

Finally, for the third set of data, the apparatus is set up with $\theta_A = 0$ and $\theta_B = t$. This is analogous to checking on the left side for a green sock and on the right side for holes. The conclusion is that $P(G, \neg H)$, the probability that the left sock (the electron at A) is green but has no holes (has the first but not the third hidden variable) is $t^2/4$.

Now it is time to put all these pieces together. We defined three hidden variables, G, S, and H and figured out a way to measure these for one electron in an entangled pair of electrons by measuring both that electron and the other one in the pair. Using the quantum mechanical properties of spin, we determined the probabilities of three specific combinations:

$$P(G, \neg S) = s^2/4 \quad P(S, \neg H) = (t - s)^2/4 \quad P(G, \neg H) = t^2/4.$$
$$(21)$$

Again, the assumption is that we can indeed speak of the electrons in an entangled pair as having the properties G, S, and H simultaneously. These would be hidden variables of the sort that EPR argued must exist. But they do not. Put the results of Equation (21) into Equation (20):

$$s^2/4 + (t - s)^2/4 \geq t^2/4. \qquad (22)$$

If indeed it is possible to speak of an electron in an entangled pair as having, or not having, properties like being green or having holes, Equation (22) must be true. But it is false! Equation (22) is equivalent to

$$t \leq s \qquad (23)$$

But we explicitly assumed that $t > s$!

Exercise 13. Get from Equation (22) to Equation (23).

The singlet state is a specific case for which Bell's Theorem, Equation (20), is not true. Therefore, we must reject the assumption that the electrons in the singlet state can have three clear binary attributes, three distinct hidden variables. The whole idea of these electrons having individual spins that point in some specific direction is invalid.

Now for the experimental verification. Although there was earlier work, Aspect, Grangier and Roger[1] did a really solid demonstration of the properties of entangled states. Their experiment was a little different from this example of Bell's Theorem, and there are a few subtle points that keep it from being a perfect proof of the theorem. Honestly, few people thought the subtle points invalidated the results, and later experiments fixed them anyway; Bell's Theorem is experimentally disproven. The probabilities of Equation (21), which contradict the probabilities of Equation (20), are the ones which actually occur.

---

[1]A. Aspect, P. Grangier and G. Roger, "Experimental Realization of Einstein-Podolsky-Rosen-Bohm Gedankenexperiment: A New Violation of Bell's Inequalities", *Phys. Rev. Lett.*, 49, 91 (1982).

Chapter 21

I Do Not Like It

So. EPR gave us an argument to the effect that there must be hidden variables. Their argument was primarily based on the fact that correlations can exist, and upon their understanding of the Uncertainty Principle for position and momentum. Bell pointed out that if there are hidden variables, Equation (20) must hold, and that in the spin-singlet state, Equation (20) does not hold. Therefore, there are no hidden variables for spin in the spin-singlet state. Why then should there be hidden variables for any other physical quantity? Experiment backs up Bell, too.

And now the perils of these waters becomes evident. Position is determined by a hidden variable; indeed, it is one of them, almost by definition, according to EPR. So if there are no hidden variables, one cannot speak of the position of an object in general. Certainly however, one can speak of the position of an object some of the time. Remember the scientist's hat? It had a position. For a certain period of time, we didn't know what the position was, but it always had one. The existence of a unique position for an object is obvious in Newtonian mechanics and elusive in quantum mechanics; yet Newtonian mechanics must obviously be a special limiting case of quantum mechanics.

It is true that EPR did use an assumption that Bell did not. EPR assumed that if an electron travels a long distance without interacting, it is possible to speak of its wavefunction independent of anything involving how it was produced. However, that is not a far-fetched assumption at all.

We cannot buy into the idea that a physical entity travels from A to B when A is measured, affecting somehow the state of B. Clearly, what is measured at B, and at A for that matter, is determined somehow at the EPR machine. But it is not possible to describe how the electron exiting the EPR machine carries whatever is determined in that machine to the measurement points A and B with some kind of hidden variable.

This is as far as we will go in these waters. The truth is that there are a host of questions and studies in this matter that would take us a long time to work through. In that sense, we really have not gone very far. Again, this is an active area of research, and a real consensus on many issues does not exist. We really can only hope to introduce you to these few central results; we cannot hope to wrap them up into a tidy bundle for you. At this point though we believe you will appreciate a quote from Erwin Schrödinger, whose understanding of, and role in the creation of, quantum mechanics is second to none. Regarding a certain attempt to assign an interpretation to the complex numbers of quantum mechanics, he famously said "I do not like it, and I am sorry I ever had anything to do with it".

Chapter 22

If You Do Not Know Who Minkowski Was, What are You Doing in His Space?

Hermann Minkowski was a mathematics professor at ETH Zurich who had a student by the name of Albert Einstein.[1] In 1905, Einstein developed the special theory of relativity[2] from the postulate that the speed of light in a vacuum, c, is a constant that does not change when we move through space at a fixed speed. That is a symmetry, actually. Here are the three steps of the symmetry:

(1) *Say something describing what we observe*: In this case, that the speed of light is measured to be some specific number of meters per second.

(2) *Transform what we are looking at in some particular way*: The transformation is to move through space at a fixed speed while looking at the light and measure its speed.

(3) *Ask if we can still say the same thing we said in step 1. If so, we have symmetry*: The final statement that the measured speed of light is exactly the same specific number of meters per second.

This is an astonishing symmetry. Suppose it was a dog instead of light. Right now there is a golden retriever taking his owner for a walk down the street at, say, three miles per hour. At step 1, we would say

---

[1] There is a legend that he once said of Einstein and relativity, "I never would have expected that student to come up with anything so clever."

[2] A. Einstein, "Zur Elektrodynamik bewegter Körper", *Annalen der Physik*, 17, 891 (1905).

that the speed of the dog is 3 mph. As we walk by in the opposite direction at four miles an hour, the speed of the dog, relative to us, is 7 miles an hour. So the number in step 3 is not 3 mph, but 7 mph. That is the way almost everything in our daily lives moves, and so it is really surprising that light is not like that.

In 1907, Minkowski[3] fully developed the idea of time as a fourth spacelike dimension. This four-dimensional spacetime is a key idea needed to develop the general theory of relativity. What we will do is go backwards, chronologically; we will start with Minkowski space, make that the central postulate of special relativity, and develop the rest of the theory from that.[4]

The full four-dimensional version of spacetime is hard to mentally picture, although it is easy to write equations about. We will enter a simplified Minkowski space with one spatial dimension, x, and one time dimension, ct. The reason for using ct instead of just t is twofold. The first, simple, reason is that we want the units of the time dimension and the space dimension to be the same, and that requires multiplying t by something that is in meters per second. The second reason is the important one. When we put ct on the same footing as x, what we are really saying is that the constant c specifies the relationship between time and space in our universe.

Figure 22.1 shows a vector in two-dimensional Minkowski spacetime. Point A is a specific place in space, x_A, as well as a specific point in time, ct_A. Imagine some object that moves from from A to B. That is, the object is at location x_A at time t_A and later is at x_B at time $t_B > t_A$. The velocity of this object happens to be a negative number in Figure 22.1; the vector corresponds to an object

---

[3]H. Minkowski, "The Relativity Principle", Lecture at meeting of Göttingen Mathematical Society 5 Nov 1907, printed in *Annalen der Physik*, 352, 927 (1915).
[4]There are many different presentations of special relativity. Probably the most succinct is Chapters 15–17 of Vol. 1 of R. P. Feynman, R. B. Leighton and M. Sands, *The Feynman Lectures on Physics* (Reading Massachusetts: Addison-Wesley, 1965). A. P. French, *Special Relativity* (New York: W.W. Norton & Co., 1968) is a full length text with conventions similar to what we use here; Supplementary Topic A of R. Resnick, *Introduction to Special Relativity* (New York: John Wiley & Sons, Inc., 1968) discusses Minkowski space in some detail.

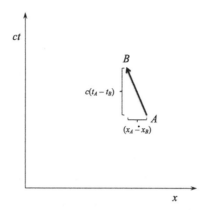

Figure 22.1: A vector from A to B in Minkowski space.

that moves towards the left as time passes and the object moves from A to B.

Be sure to distinguish between the ct that we draw on the page and the passage of time which we, as humans, experience. In Figure 22.1 we have space and time laid out on a plane and see all of it at once in some sort of transcendental way. This ignores the passage of time that we perceive. That experiential time is not quite the time that we draw on the vertical axis of Figure 22.1. When we bolt ct to x to create spacetime, we implicitly ignore that for us as humans, the past, present, and future are quite different. One is remembered, one is fleeting, and one is unknown to us all. On Figure 22.1 though, this is all just some other direction on the plot.

Minkowski space is what the mathematicians call a hyperbolic space. That is, the "distance" between A and B now has a minus sign in it. We are used to the sum of the squares of the two sides being the hypotenuse. In a hyperbolic space, the *difference* of the squares of the two sides is the side of the hypotenuse! Like so:

$$s^2 = (x_A - x_B)^2 - (ct_A - ct_B)^2 \tag{24}$$

and not like so:

$$s^2 = (x_A - x_B)^2 + (y_A - y_B)^2. \tag{25}$$

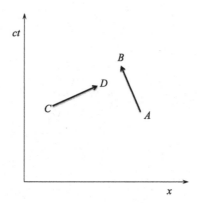

Figure 22.2: Two vectors in spacetime that appear to have the same length but do not.

That minus sign changes everything! All the bizzare effects of the special theory of relativity come from the minus sign in Equation (24). What remains is to work out what they are.

The minus sign can make it a hard to picture distances in Minkowski space. In Figure 22.2, we show two vectors, AB and CD. To the eye, they are of the same length. However, because they are in different directions, they have different lengths as measured using Equation (24). In fact, while AB has $(ct_B - ct_A)^2 > (x_B - x_A)^2$, and so s^2 is negative for AB, CD has $(ct_D - ct_C)^2 < (x_D - x_C)^2$, and so s^2 is positive for CD.

Exercise 14. If the vector CD had $s = 0$ exactly, what direction would it point in?

In normal Euclidian space, the length of a vector, as given by Equation (25), is not changed by drawing different coordinates. In particular, if the coordinate system is rotated through some angle, the length s is not changed. In Minkowski space, with its minus sign, that same thing is true, nearly. The distance s given by Equation (24) is unchanged if we, um, sort-of-rotate.

Before we get into the sort-of-rotation that keeps the length s^2 of Equation (24) unchanged, let us talk about the ordinary rotations that keep the length s^2 of Equation (25) unchanged.

Chapter 23

Rotational Symmetries and Matrices

Figure 23.1 shows a vector in two rotated coordinate systems, using a plain old ordinary Euclidian geometry.

Here are the three steps of the symmetry for Figure 23.1:

(1) *Say something describing what we observe*: In this case, we claim that the distance from A to B is, in the unprimed frame, given by Equation (25).

(2) *Transform what we are looking at in some particular way*: The transformation is to rotate the coordinates, creating the primed coordinate system of Figure 23.1.

(3) *Ask if we can still say the same thing we said in step 1. If so, we have symmetry*: The final statement is that the length in the primed frame is the same as in the unprimed frame.

With that proven, then we then say that the length of a vector is symmetric with respect to rotations.

The distance from A to B using the rotated system, also called the primed system, is s', where

$$s'^2 = (x'_A - x'_B)^2 + (y'_A - y'_B)^2. \tag{26}$$

The existence of the symmetry is proved if we can prove that $s = s'$. To do that, we need to be able to connect the (x, y) coordinates to the (x', y') coordinates. The way to do that is with

$$x'_A = \cos(\phi)x_A + \sin(\phi)y_A$$

$$y'_A = -\sin(\phi)x_A + \cos(\phi)y_A$$

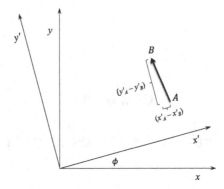

Figure 23.1: A vector in Euclidian space, in two coordinate systems. The (x', y') primed frame, is at an angle ϕ relative to the (x, y) unprimed frame.

and similarly

$$x'_B = \cos(\phi)x_B + \sin(\phi)y_B$$
$$y'_B = -\sin(\phi)x_B + \cos(\phi)y_B. \tag{27}$$

To see that Equations (27) are indeed correct, pick a point $(x, y) = (1, 0)$ and draw it on Figure 23.1, and then figure out what the (x', y') coordinates of that point are. Repeat, but with $(x, y) = (0, 1)$.

Exercise 15. Take the expressions for x'_A, y'_A, x'_B and y'_B from Equations (27) and put them into Equation (26). Show that $s'^2 = s^2$ as given in Equation (25); hence, $s' = s$.

Equations (27) are, in matrix[1] form,

$$\begin{bmatrix} x' \\ y' \end{bmatrix} = \begin{bmatrix} \cos(\phi) & \sin(\phi) \\ -\sin(\phi) & \cos(\phi) \end{bmatrix} \begin{bmatrix} x \\ y \end{bmatrix}. \tag{28}$$

---

[1]We will do little with matrices besides multiply and invert them. There are several good introductory texts; D. Lay, S. Lay and J. McDonald, *Linear Algebra and Its Applications* (New York, London: Pearson Education, Inc., 5th Ed., 2016); B. Kolman and D. R. Hill, *Elementary Linear Algebra with Applications* (New York, London: Pearson Education, Inc., 9th Classic Edition, 2008; 9th Ed., 2018) and G. Strang, *Introduction to Linear Algebra* (Wellesley MA: Wellesley-Cambridge Press, 4th Ed., 2009; 5th Ed., 2016) all cover this material in their opening chapters and go on to reveal the detailed and immensely useful world of linear algebra.

As we discussed back in Groups, rotations through an angle form a group, the circle group. Here, we are rotating a coordinate system through the angle ϕ. The matrix of Equation (28) represents a rotation of a coordinate system through an angle ϕ; it must be that the set of all matrices of this form are a group. This group is called $SO(2)$.

A group, as you recall, is a set with an operation. In this case, the operation will be matrix multiplication. The operation, with that set, must have the properties of closure, associativity, the existence of an identity element, and invertibility. We will go through all four of these properties for this group in detail because it that process will tell us something useful about groups made of square matrices.

Regarding closure, for two matrices A and B which are of the form in Equation (28), the product AB is also of the form in Equation (28). If

$$A = \begin{bmatrix} \cos(\phi) & \sin(\phi) \\ -\sin(\phi) & \cos(\phi) \end{bmatrix} \quad \text{and} \quad B = \begin{bmatrix} \cos(\theta) & \sin(\theta) \\ -\sin(\theta) & \cos(\theta) \end{bmatrix} \quad (29)$$

then

$$AB = \begin{bmatrix} \cos(\phi + \theta) & \sin(\theta + \phi) \\ -\sin(\theta + \phi) & \cos(\phi + \theta) \end{bmatrix} \quad (30)$$

which is also of the form in Equation (28).

Regarding associativity, the requirement is that for any three matrices A, B, and C, we must have that $(AB)C = A(BC)$. This requirement is met by the definition of matrix multiplication. However, the proof is detailed enough that we banish it to the back of the book.

Exercise 16. For the square matrices A, B, and C, show that $(AB)C = A(BC)$.

The identity element for $SO(2)$ is just

$$\begin{bmatrix} 1 & 0 \\ 0 & 1 \end{bmatrix}. \quad (31)$$

And for any group of square matrices, there is an identity element, which is all zeros with ones on the diagonal.

For invertibility, matrix multiplication does provide a definition of an inverse for us, but it is not always true that the inverse for any given matrix exists. For a 2×2 matrix, the inverse is given by

$$\begin{bmatrix} a & b \\ c & d \end{bmatrix}^{-1} = \frac{1}{ad - bc} \begin{bmatrix} d & -b \\ -c & a \end{bmatrix} \tag{32}$$

and the expression $ad-bc$ is the determinant. Determinants exist for square matrices of any size, and a square matrix can only be inverted if its determinant is zero. So to show that a set of square matrices is a group, we need to prove that no element in the group has a determinant of zero.

Exercise 17. Show that for matrices of the form in Equation (28), the determinant is not zero.

So we have shown that the set of square 2×2 matrices of the form in Equation (28) does indeed form a group. Along the way, we have discovered that any set of square matrices can be proven to be a group by checking only that we have closure and a nonzero determinant. That simplifies the problem of determining if a specific set of matrices is a group.

Chapter 24

The Sort-of Rotation

Figure 24.1 will help us to realize what a sort-of-rotation in Minkowski space really is. Again, imagine some object that moves from A to B. In the original coordinates, the velocity of this object is still a negative number, just as in Figure 22.1; the vector corresponds to an object that moves toward a smaller value of x as time passes and the object moves from A to B. In other words, the x value at A is bigger than the x value at B.

But now, in Figure 24.1, we have added a ct' coordinate and axis. There has to be an x' coordinate to go with it, but we will come to that later. We picked the direction of the ct' axis to be exactly in the same direction as the vector from A to B, so that in the new set of coordinates the motion from A to B will be in the time direction only. In the new frame, going from A to B involves no spatial motion at all.

What does the new coordinate system mean, physically? Each coordinate system is a frame of reference in the sense that we can imagine a person, an observer, on the time axis. As that person's watch ticks, he proceeds from one point on the ct axis to another point higher up on that axis. For that observer, an object at any particular x is a distance x meters away. The observer for the first, unprimed ct frame of reference — let us call him Urie — proceeds straight towards the top of the page as his watch ticks. The new, primed, ct' axis in Figure 24.1 corresponds to a second observer — call her Petra — moving relative to Urie at a steady speed. (The steady speed part is important — more on that when we come

101

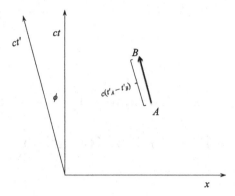

Figure 24.1: A spacetime interval in two different coordinate systems that are boosted relative to each other. The x' axis is not drawn yet.

to Homer and O.Th.R. Sock.) From Urie's point of view, in the unprimed frame of reference, as his watch ticks and he moves up the ct axis, the object going from A to B is moving towards him, and Petra is moving away from him. Both motions, from Urie's point of view, are in the $-x$ direction. From Petra's point of view, in the primed frame of reference, the object going from A to B is not moving at all, neither towards or away from her. However, from Petra's point of view, Urie is moving away from Petra, in the $+x'$ direction.

Exercise 18. By drawing a line perpendicular to the ct axis, show that $\tan(\phi) = v/c$, where v is the speed of Petra, the second observer, as seen by Urie, the first observer.

Here is a physical example corresponding Figure 24.1. Urie puts a ruler on the floor of a train car, with the zero next to his seat and the seat in front of him at, say, $+1$ m. The next seat in front of him is at $+2$ m, the one in back of him is at -1 m. That is the unprimed frame of reference. Petra is standing on the platform as the train rolls through at a steady velocity, facing the train and holding a ball off to her side. She has a negative velocity relative to Urie on the train; her x position steadily and smoothly gets lower and more negative. He is at rest in the unprimed frame of reference. The ball, which is not moving from Petra's point of view, also has a negative velocity from Urie's point of view, like the vector AB.

If two coordinate systems are sort-of-rotated relative to each other in Minkowski space, objects that are at rest in one coordinate system are moving uniformly at some velocity (we say "are boosted") in the other.

We will want to have a couple of handy variables, β and γ. The definitions are $\beta = v/c$, where v is some specific velocity, and $\gamma = 1/\sqrt{1 - \beta^2}$. Since velocity can be either positive or negative, but nothing can go faster than c, it follows that $|\beta| \leq 1$. That means in turn that $\gamma \geq 1$; when an object is at rest and has no velocity, $\beta = 0$ and $\gamma = 1$.

We wrote a matrix in Equation (28) to show how coordinates change in ordinary space when the axes are rotated through an angle ϕ. A similar matrix exists for a boost, which is our sort-of-rotation in Minkowski space. We can use that matrix to draw the x' axis. The boost matrix is

$$\begin{bmatrix} x' \\ ct' \end{bmatrix} = \begin{bmatrix} \gamma & \beta\gamma \\ \beta\gamma & \gamma \end{bmatrix} \begin{bmatrix} x \\ ct \end{bmatrix}. \tag{33}$$

Exercise 19. Prove Equation (33).

Let us look at the boost matrix a little more closely. Because $\gamma = 1/\sqrt{1 - \beta^2}$, the whole matrix is determined by a single variable, β. That is not a big surprise; a boost matrix represents motion at some fixed constant velocity and that velocity can be expressed with a single number for a single spatial dimension.

The second thing to notice is that the upper right and lower left elements of the boost matrix have the same sign, which is not like a rotation matrix in normal space at all. In Equation (28), the upper right and lower left elements have different signs. This is why we did not draw an x' axis in Figure 24.1. We are not in normal space, so our intuitive tendency to draw the x' axis at a right angle to the ct' axis would not work for Minkowski space.

A final very important thing to notice is that the length of a vector in Minkowski space is unchanged by a boost, just as the length of a vector in Euclidian space is not changed by a rotation. The equivalent of Exercise 15 is

Exercise 20. Show that Equation (33) implies that $(x')^2 - (ct')^2 = (x)^2 - (ct)^2$; the length of a vector in Minkowski space is unchanged by applying a boost.

To find the direction of the x' axis, we try to find a vector in the unprimed coordinates that has $x' = 1$ and $ct' = 0$ in the primed coordinates. That is the unit length direction vector for the x' direction, and we want its location in the unprimed coordinates in order to know where to draw it. In other words, we want to solve

$$\begin{bmatrix} 1 \\ 0 \end{bmatrix} = \begin{bmatrix} \gamma & \beta\gamma \\ \beta\gamma & \gamma \end{bmatrix} \begin{bmatrix} x \\ ct \end{bmatrix} \tag{34}$$

for x and ct. Using Equation (32), the result is

$$\begin{bmatrix} x \\ ct \end{bmatrix} = \begin{bmatrix} \gamma \\ -\beta\gamma \end{bmatrix}. \tag{35}$$

Exercise 21. Get from Equation (34) to Equation (35) by finding and applying the inverse of the boost matrix in Equation (34). Along the way, show that the boost matrix of Equations (34) and (33) and its inverse both have determinant 1.

When we draw the x' axis in the unprimed coordinates, it has negative ct! The point defined in Equation (35) is the dot on Figure 24.2. The angle between the x and x' axes is the same as the one between the ct and ct' axes — it is $\tan^{-1}(\beta)$.

Exercise 22. Figure 24.2 shows how to move from a coordinate system in which an object has a negative velocity to a coordinate system in which it is at rest. Draw the same figure, but for an object initially with a positive velocity.

Figure 24.2 and Equation (33) are in effect, rotations in Minkowski space. In fact, with some different conventions from what we have used here, Equation (33) can be made to very much like Equation (28). These boosts have many of the properties of rotations in normal space. Like rotation matrices, they are defined by a single parameter. They also have determinants of one. The product of any

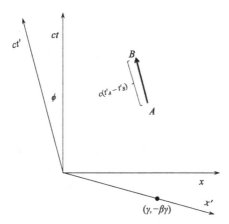

Figure 24.2: A spacetime interval in two different coordinate systems that are boosted relative to each other. The x' axis is now drawn.

two boost matrices is also a boost matrix, providing the closure property. That means that the boost matrices form a group.

The product of any two boost matrices is also a boost matrix. If we start with some object at rest and apply a boost, that is the same as having the object move at a constant velocity. An example is... you hold a ball in your hand as you stand on the platform of a train station. Your friend in the train reaches out and grabs the ball as the train rolls through the station at some velocity v. Now the ball, from your point of view, has a boost with $\beta_1 = v/c$. Suppose then that your friend throws the ball towards the front of the train with a velocity u; so the second boost, relative to the first, is $\beta_2 = u/c$. Now the ball is still moving with some velocity from your point of view as you stand there on the platform, so there must be some boost matrix for the combined velocity. So we expect that if you apply a boost matrix to some vector in Minkowski space, and then apply another boost matrix to the result, we should get some kind of combined, valid, boost matrix.

The way to write two boosts applied in sequence in Minkowski space is,

$$
\begin{bmatrix} \gamma_2 & \beta_2\gamma_2 \\ \beta_2\gamma_2 & \gamma_2 \end{bmatrix} \left(\begin{bmatrix} \gamma_1 & \beta_1\gamma_1 \\ \beta_1\gamma_1 & \gamma_1 \end{bmatrix} \begin{bmatrix} some \\ place \end{bmatrix} \right) \tag{36}
$$

where *some place* is some arbitrary vector in Minkoski space and the two boost matrices are applied sequentially. What we expect is that Equation (36) should be the same as some combined boost with combined $\gamma = \gamma_T$ and $\beta = \beta_T$, like so:

$$\begin{bmatrix} \gamma_T & \beta_T \gamma_T \\ \beta_T \gamma_T & \gamma_T \end{bmatrix} \begin{bmatrix} some \\ place \end{bmatrix}. \tag{37}$$

That is, we expect on physical grounds that the boost matrices to have the closure property that makes boosts in Minkoski space a group, just like rotations in regular space. Indeed that is so. Here is the key result:

$$\beta_T = \frac{\beta_1 + \beta_2}{1 + \beta_1 \beta_2}. \tag{38}$$

Exercise 23. Derive Equation (38).

Equation (38) contains a surprise. It tells us that we can not go faster than the speed of light.

Chapter 25

299,792,548 Meters per Second — and No More!

To unwrap Equation (38), let us go back to our example of handing a ball from a platform to a moving train and then throwing it towards the front train. Say that the first train is moving past the platform at 3 meters per second. Then $\beta_1 = (3\,\text{m/s})/(3\times10^8\,\text{m/s} = 1\times10^{-8}$. Just to keep things simple, we will say that the speed of light is exactly $3 \times 10^8\,\text{m/s}$. And let us say that your friend throws the ball forward at the same velocity, $3\,\text{m/s}$. That makes $\beta_2 = \beta_1 = 1 \times 10^{-8}$. How fast is the thrown ball moving relative to the platform?

A stranger to Minkowski space would expect that the answer is $(3\,\text{m/s}) + (3\,\text{m/s}) = 6\,\text{m/s}$, or equivalently, $\beta_T = \beta_1 + \beta_2 = 2 \times 10^{-8}$; in other words that velocities add just like that. Actually, from Equation (38), the correct result is $\beta_T = 2(1 \times 10^{-8})/(1 + 10^{-16})$. That is extremely close to the simple sum, to 2×10^{-8}, because the denominator is so close to 1. The denominator is close to 1 as long as β_1 and β_2 are small. Of course most things in our daily life are moving slowly compared to light, and that is why Einstein's theory is so astonishing.

As we start to deal with objects that are faster, things start to change because the denominator in Equation (38) begins to get larger, eventually reaching a maximum value of 2. Instead of adding two velocities of $\beta = 10^{-8}$, try combining two velocities of $\beta = 10^{-1}$; that is, $3 \times 10^7\,\text{m/s}$. Then $\beta_T = 2(1 \times 10^{-1})/(1 + 10^{-2})$. That differs from simple addition by 1%; still not a very big effect. But at $\beta = 0.9$, we have $\beta_T = 2(0.9)/(1+0.81) = 0.994$, which is far from the simple addition of $0.9 + 0.9 = 1.8$.

The claim that one cannot go faster than the speed of light is the claim that β_T, regarded as a function of β_1 and β_2, i.e., $\beta_T = \beta_T(\beta_1, \beta_2)$ never rises above 1.

Exercise 24. Show that $\beta_T = \beta_T(\beta_1, \beta_2) = (\beta_1 + \beta_2)/(1 + \beta_1\beta_2) \leq 1$.

Chapter 26

Going Slower by Going Faster

In the time it takes you to read this sentence, a muon probably went through your head. It came from probably 15 or 20 kilometers up in the atmosphere, where it was created when some particle from outer space, called a cosmic ray, hit a molecule, probably of nitrogen or oxygen. This happens all the time. In any given second, a couple dozen or so muons go through you, depending on the your altitude; there are more cosmic rays on mountain tops than at sea level.

The muons are fast; they are travelling at pretty close to the speed of light. That means it takes them some 50 or 70 microseconds to get to your head from the top of the atmosphere. But if you have a muon in your hand it will disappear in a fraction of that time — about 1.5 or $2\,\mu$s; it decays into an electron and a couple of antineutrinos. How does the muon survive the time it takes to travel to the surface of Earth from up in the atmosphere?

What is happening is that time goes slower for the muons because they are moving at close to the speed of light. Start with the muon in your hand; it is at rest, meaning that for the $2\,\mu$s while it is in your hand, the position x does not change. In Minkowski space, its position does not change, and the muon travels only in time. The Minkowski space vector for the muon in your hand is

$$\begin{bmatrix} 0 \\ c(2\,\mu\text{s}) \end{bmatrix} \tag{39}$$

and if we look at that muon when it is moving very fast, it has a boost applied to it (Equation (33)):

$$\begin{bmatrix} \gamma & \beta\gamma \\ \beta\gamma & \gamma \end{bmatrix} \begin{bmatrix} 0 \\ c(2\,\mu s) \end{bmatrix} = \begin{bmatrix} \beta\gamma c(2\,\mu s) \\ \gamma c(2\,\mu s) \end{bmatrix}. \tag{40}$$

What does that mean? The number on the top, $\beta\gamma c(2\,\mu s)$, is the distance that the muon travels before it decays. The number on the bottom, $\gamma c(2\,\mu s)$, is c by the amount of time before it decays. The decay time is no longer $2\,\mu s$, but rather is $2\,\mu s$ times γ; and γ is a big number. Because the muon is travelling quickly, it decays more slowly. Time travels more slowly for rapidly moving objects.

Exercise 25. A typical value of β for a muon in cosmic rays is $\beta = 0.999651076$. What is γ? What is the $2\,\mu s$ decay time stretched out to? How far does such a muon travel in that stretched out time?

More generally, if instead of the $2\,\mu s$ lifetime of the muon we consider any interval of time Δt, that interval takes $\gamma\Delta t$ if we look at it with a boost.

This effect is called time dilation; because of it, if we could construct spaceships that travel at very high speeds and go to other planets, then the astronauts on those spaceships would age more slowly and could return to Earth abnormally young.

Exercise 26. Which member of the rock band Queen has a PhD in astrophysics? What is the song '39 from the 1975 album "A Night At The Opera" about?

Chapter 27

The Twins

Mr. Homer Sock and his brother, Oswald Theodore Reginald (O.Th.R.) Sock are twins, but not quite identical twins. There is one crucial difference between the two of them. O.Th.R. Sock, like so many Socks, has a deep-seated case of wanderlust, but Homer utterly dislikes travel. In the year 2139, O.Th.R. boards a rocket ship that travels at 99.98% of the speed of light[1] to the planet There, which is 50 light-years from planet Earth. He immediately turns around and returns to Earth, where he has a joyous reunion with his brother Homer, who stayed at home the whole time.

At $\beta = 0.9998$, $\gamma = 50$. From Homer's point of view it would take 50 years for O.Th.R. to get to There at the speed of light, and actually, O.Th.R. is traveling at pretty darn close to that speed. To be precise, from Homer's point of view it takes $50/0.9998$ years for O.Th.R. to get to There, and the same time to come back. That works out to about 100 years and one week, total. Homer will see his brother in the year 2239.

But time passes more slowly for O.Th.R., by a factor of 50. He does not age by 50 years in the time it takes him to get to There; he ages only 1 year. O.Th.R. is only two years older when he sees Homer again.[2]

---

[1] The calculations here assume that $\gamma = 50$ exactly, and that the distance to There is 50 light years exactly; so for example, $\beta = 0.997998$. We also take a year to be 365.24 days.

[2] By the way, O.Th.R., seeing that his trip to There requires only a single year, concludes that the distance to from Earth to There is only 1 light year, not 50; this is known as Lorentz contraction.

But... wait. From the point of view of O.Th.R., he is stationary and Homer is the one who moves, first away, and then back. From O.Th.R.'s point of view, time moves slower for Homer, does it not? From O.Th.R.'s point of view, time for Homer moves slower by a factor of 50 for the two years when O.Th.R. is travelling, meaning that Homer has aged by only 2 years/γ = 14.6 days when his brother returns. So which is it? When the brothers reunite, is Homer a century older than when they separated? Or is he only two weeks older?

The correct answer is that Homer is indeed a century older. The short reason is that we broke an assumption when O.Th.R. turned around his rocket ship. All the discussion to this point applies to objects moving past each other at fixed velocities. We have not at all discussed what happens when O.Th.R.'s velocity changes as he decelerates to a stop at There and then turns around and starts to accelerate back homeward.

Looking more carefully at what happens when O.Th.R. turns around will lead us to the doorstep of the general relativity. We will not actually go in, but we can take a look around.

Start with the Minkowski space diagram for the moment after O.Th.R. and Homer part ways, Figure 27.1. That Figure uses the result of Exercise 22 to determine the ct and x axes for O.Th.R.; his spaceship, is moving in the $+x$ direction with respect to Earth, and the angles of O.Th.R.'s spaceship's coordinate system become acute angles rather than the obtuse angles of Figure 24.2.

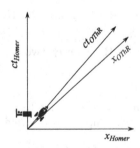

Figure 27.1: Minkowski space diagram for shortly after O.Th.R.'s spaceship's liftoff.

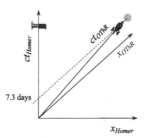

Figure 27.2: Minkowski space diagram for shortly before O.Th.R.'s space ship reaches There.

Figure 27.2 shows the situation just shortly before O.Th.R. reaches the rather greyish planet There. He has traveled c (1 year) along his ct axis. The dashed line parallel to the X_{OThR} axis, marks all the places where in spacetime, from O.Th.R.'s point of view, the time is the same as that which he reads on his watch. The point where that dashed line intersects Homer's time axis is the point in Minkowski space on Earth where, from O.Th.R.'s point of view, one year has passed. On earth, one week has passed for Homer.

Shortly after O.Th.R. turns around, the situation is different. His boosted frame of reference is shown in Figure 27.3. The intersection of his new x_{OThR} axis with Homer's timeline is the point where Homer is, from O.Th.R.'s point of view, at the same time as O.Th.R. At that point, Homer is suddenly much older — almost a century older! In O.Th.R.'s frame of references, Homer has aged almost a century, and this aging all happened while he was turning around his spaceship.

Homer never accelerated. His coordinate system never changed, and he always sees his twin aging at a uniform and steady rate. There is no symmetry between Homer's and O.Th.R.'s experience.

In turning his ship around, O.Th.R. accelerated. Plainly, acceleration changes the rate at which time flows with effects that are above and beyond the time dilation that we have already discussed. So far, we have been discussing only the special theory, which is for the special case where there is no acceleration.

As Einstein realized in 1907, acceleration and gravity are closely linked. The general theory of relativity begins by noticing that under the force of gravity, all objects move in the same way. That is what

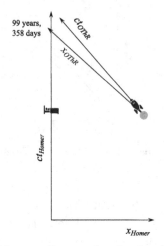

Figure 27.3: Minkowski space diagram for shortly after O.Th.R.'s space ship leaves There.

Galileo proved in Pisa! This is also what happens when a frame of reference accelerates. Suppose you are sitting in a windowless spaceship, and you find yourself pushed down in your seat with a force twice what you normally weigh. Are you on a planet where the force of gravity is twice what it is on earth? Or are you accelerating upwards at a rate twice as fast as objects on earth fall? The truth is that you cannot really tell. This symmetry, this equivalence between acceleration and gravity, is what underlies the general theory of relativity.

We will not get into the general theory really, but we can give you a flavor of what happens. In normal space, the distance between two points can be written

$$s^2 = [x \; y] \begin{bmatrix} 1 & 0 \\ 0 & 1 \end{bmatrix} \begin{bmatrix} x \\ y \end{bmatrix}. \tag{41}$$

That is just the Pythagorean theorem, written in matrix form. In Minkowski space, the equivalent is

$$s^2 = [x \; ct] \begin{bmatrix} 1 & 0 \\ 0 & -1 \end{bmatrix} \begin{bmatrix} x \\ ct \end{bmatrix}. \tag{42}$$

In the general theory of relativity, that matrix in the middle becomes all kinds of interesting things.[3] Imagine a point trapped inside a curved two-dimensional surface like a sphere or a potato chip, and try to compute the distance it travels going from one point to another. That requires an equation like Equation (42) but with some nondiagonal matrix in the middle. It is in this sense that gravity can be thought of as spacetime being curved.

---

[3] Actually it becomes all kinds of interesting symmetric things. The matrix in the middle must equal its transpose. Towards the end of his life, Einstein studied the case of an antisymmetric matrix which is equal to -1 times its transpose in an unsuccessful effort to theoretically unify the gravitational and electromagnetic forces.

Chapter 28

Momentum in Minkowski Space

To show that $E = mc^2$, we will need the concept of momentum in Minkowski space.

Going back to the days of Newton, we have known of momentum as the mass, m, of some object, times the object's velocity. The mass is a measure of the amount of inertia that an object has;[1] it is a scalar, even in Minkowski space. A velocity is a change in position divided by a change in time. In Euclidean space, the velocity vector is

$$\vec{v} = \begin{bmatrix} \Delta x \\ \Delta y \end{bmatrix} \div \Delta t = \begin{bmatrix} \Delta x/\Delta t \\ \Delta y/\Delta t \end{bmatrix} \tag{43}$$

where the vector of Δx and Δy is the change in position that happens during the time interval Δt.

In Minkowski space we might try then to define the velocity with

$$\begin{bmatrix} \Delta x \\ c\Delta t \end{bmatrix} \div \Delta t = \begin{bmatrix} \Delta x/\Delta t \\ c \end{bmatrix} \tag{44}$$

Such a quantity can be defined, but now that we know about time dilation, we can realize that it is not of much use. Time dilation means that Δt is not the same in every coordinate system; and because we understand now the importance of symmetry, we realize that interesting things will be the same in every coordinate system. What we want is to divide by something that is the same in every

---

[1] The mass is also a measure of how much an object creates and responds to gravity, but that is not the aspect of mass that is important here.

coordinate system. That something also has to equal Δt for objects moving at low speeds. If it does not, then our new definition of velocity is different than the standard Newtonian one for objects moving at much less than the speed of light.

If an object with no velocity moves from point A to point B in Minkowski space, then point B is directly above point A in our diagrams, and the time interval between A and B is, using Equation (24), is $\Delta \tau^2 = -s^2/c^2$. Now s is the same in every coordinate system, regardless of what boost has been applied. What we want is s, but rearranged somehow to look like Δt. Define the proper time, $\Delta \tau$, like so:

$$\Delta \tau^2 \equiv -s^2/c^2 = \Delta t^2 - \Delta x^2/c^2 = \Delta t^2[1 - (\Delta x^2/\Delta t^2)/c^2]. \quad (45)$$

The quantity $\Delta x/\Delta t$ is our old friend v. Let's keep that around and continue to call it the velocity of the object, or, when divided by c, the boost. Let's use $\Delta \tau$, the change in proper time, rather than Δt, the change in time, in trying to construct a velocity for Minkowski space. Since

$$\Delta \tau = \Delta t \sqrt{1 - \beta^2} = \Delta t/\gamma, \quad (46)$$

the good way to define the velocity of an object in Minkowski space is

$$v^{(M)} \equiv \begin{bmatrix} \Delta x \\ c\Delta t \end{bmatrix} \div \Delta \tau = \gamma \begin{bmatrix} \Delta x \\ c\Delta t \end{bmatrix} \div \Delta t = \gamma \begin{bmatrix} v \\ c \end{bmatrix}. \quad (47)$$

Because momentum is velocity times inertia, the next step is to multiply this by the inertia, m. The magnitude of $v^{(M)}$ is the same before and after any boost. Therefore, the velocity in Minkowski space times m is also not going to change its value in different coordinates. That is the definition of an object's momentum in Minkowski space: $p^{(M)} \equiv mv^{(M)}$:

$$p^{(M)} = m \begin{bmatrix} \gamma v \\ \gamma c \end{bmatrix}. \quad (48)$$

Chapter 29

Why E Is In Fact mc^2

Now it is possible to prove the famous equation.

Consider some object with some inertia m moving with some velocity v. The spacelike component of Equation (48) is γmv. When the object's velocity is small, $\gamma \approx 1$ and therefore $\gamma mv \approx mv$, which is what we thought the momentum was before we went into Minkowski space.

Let us take a moment and ponder this. In Newtonian mechanics, the momentum is $p = mv$. From Noether's theorem, we have the concept that space is to time as momentum is to energy. We wrote a vector in Minkowski space, Equation (48), that we designed to be the momentum. In its space component, there is a quantity γmv which reduces to the Newtonian momentum at speeds much less than the speed of light; that is the relativistic momentum. At large velocities, there is an extra factor of γ in the momentum to allow for relativistic effects. Immediately we wonder if, given the analogy we learned from Noether's theorem, the time like component of Equation (48), γmc, is the energy of a moving particle.

That timelike component, at slow speeds when β is small, is

$$\gamma mc = mc/\sqrt{1 - \beta^2} \cong mc(1 + \beta^2/2 + 3\beta^4/8 + \cdots) \qquad (49)$$

where the terms of order β^4 and higher are small and can be ignored.

Exercise 27. Using the binomial theorem, check that $1/\sqrt{1 - \beta^2}$ is indeed approximately equal to $(1 + \beta^2/2)$ for low β.

So for velocities that are small, for slow moving objects,

$$\gamma mc \cong mc + mv^2/(2c). \tag{50}$$

Remember that Isaac Newton taught us that the energy of a moving object is $\frac{1}{2}mv^2$. That is c times the second term in Equation (50). That tells us that the timelike coordinate of the momentum in Minkowski space is E/c. Multiply both sides of Equation (50) by c to get

$$E = \gamma mc^2 \cong \frac{1}{2}mv^2 + mc^2. \tag{51}$$

The energy is the familiar $\frac{1}{2}mv^2$ plus another term, mc^2, which is some energy that an object has just as a result of it having inertia. And now we can rewrite Equation (48), to get

$$p^{(M)} = \gamma m \begin{bmatrix} v \\ c \end{bmatrix} = \begin{bmatrix} p \\ E/c \end{bmatrix}. \tag{52}$$

Let us work out a few more consequences of Equation (52). Remember that the length of a vector, in either Euclidian or Minkowski space, is invariant. It has the symmetry property that it does not change when it is boosted. The length *squared* of the momentum vector in Minkowski space is some constant quantity

$$p^2 - (E/c)^2. \tag{53}$$

When the object is at rest, the momentum p is zero and the energy E is mc^2. That means the fixed number is just $-(mc)^2$.

$$-(mc)^2 = p^2 - (E/c)^2$$

or

$$E^2 = (mc^2)^2 + (cp)^2. \tag{54}$$

That is the full form of $E = mc^2$. In addition to the mc^2 term, there is another term, which depends on the momentum, for the kinetic energy.

Exercise 28. Considering the spacelike and timelike elements of Equation (52), show that $\beta = cp/E$. Use this last equation to eliminate E in Equation (54) and discover that an object travelling exactly at the speed of light cannot have mass.

Chapter 30

Antimatter

Equation (54) is a formula for E^2, not for E. To get the energy, we need to take the square root; and there is both a positive answer and a negative one.

What does it mean, negative energy? When thinking of kinetic energy, the idea makes no sense; an object that is moving has positive kinetic energy, and an object that is not moving has zero kinetic energy; negative kinetic energy is impossible. Regarding mc^2, negative energy also makes no sense; it would mean that the mass is negative, but negative inertia is nonsense. It is true that one can have a negative potential energy, but that is just a mathematical oddity that can easily be made to disappear. The point in space where the potential energy is set equal to zero can be selected arbitrarily; it is only changes in potential energy that matter.

For a long time nobody really thought this was a problem. All we have done in Equation (54) is compute the length of a vector in Minkowski space. Just as when finding the length of a vector in ordinary space, you can just ignore the negative answer.

In 1926 Klein and Gordon tried to get the quantum mechanical form of Equation (54), and at this point the negative energy solutions became a problem. The probability for a particle to be at a specific position using the Klein–Gordon equation is proportional to the energy. A negative energy corresponds to a negative probability, which makes no more sense than negative energy. Worse, a complete set of both positive and negative energy solutions are needed in order to solve any practical problem with the Klein–Gordon equation.

In 1928, Paul Dirac, thinking that the problem was that there is an E^2 instead of just an E in Equation (54), decided to work out what would replace the Klein–Gordon equation if he allowed only E, not E^2, in some equation relating energy to momentum for quantum motion. He still wound up with negative energy solutions. They simply do not go away! In fact, the situation is worse. For the Klein–Gordon equation, there are separate positive and negative energy solutions. For the Dirac equation, a single solution has both positive and negative energy parts, wrapped up together. Basically, every solution has negative energy components.

Imagine a positively charged donkey and a negatively charged carrot. Well, opposites attract. The donkey will be pulled towards the carrot and you could get energy out, perhaps in the form of getting a cart pulled. On the other hand, like charges repel. Imagine a negatively charged donkey and a negatively charged carrot. Now one must push the equine to get it to move, and that involves putting energy into the system.

So changing the electrical charge is the same as changing the sign of the energy. And that is what antimatter is; it is the negative energy solution, which is the same as a particle with opposite electrical charge.

That is for electrical, and of course also magnetic, forces. There are other kind of forces too. We've already mentioned the weak force, and the strong force. And of course, there is gravity.

In the case of the strong force, the single kind of charge that appears in electromagnetic forces (positive vs. negative) is replaced with three kinds of charges. There is the red charge (red vs. antired), the green charge (green vs. antigreen) and the blue charge (blue vs. antiblue). But it is still true that the antiparticles have the anticharge.

Gravity is different. Remember Galileo at the Leaning Tower of Pisa? All objects respond to gravity by moving in the same way. That is the basic fact that goes into the general theory of relativity. It is not possible to have two objects that are otherwise the same that move differently because of gravity. This is quite unlike electromagnetic or strong forces, where objects that are otherwise the same could have different charges and therefore move differently. In this sense, there is no such thing as a gravitational charge.

For this and other reasons, physicists expect that antimatter falls down, just like matter, and does not fall up. But nobody has yet an experiment designed that will get a clear verification of this expectation, so it is an open question.

As for weak forces, the situation is more complicated. There is a charge, but often, in the place where a charge would go into the calculations, there also is a 3×3 matrix of complex numbers. As a result, there are differences between particles and antiparticles in terms of how they interact or decay. That was the conclusion of the discovery of Christenson, Cronin, Fitch, and Turley that we mentioned in A Symmetry That Is Not.

Not all elementary particles have antimatter forms. The photon, which is a particle of electric and magnetic field, does not have an antiphoton, and neither do the corresponding particles for the strong force. In the case of the weak force, there are three particles that carry the force; the neutral one does not have an antiparticle form. When a particle and its antiparticle collide, they annihilate into these antimatterless particles.

Chapter 31

Your First Nuclear Physics Theory: Protons and Neutrons

In the early 1930s it was known that there were protons and neutrons inside the nucleus of the atom. It was also known that protons and neutrons had very similar masses. The mass difference between the proton and the neutron is quite small — 1 part in 700 or so. And the forces between two neutrons, or between two protons, or between a neutron and a proton, are the same. Both the neutron and proton are very different in mass from the electron and its antiparticle, the positron, the other particles that were known at that time. The proton has an electric charge, and the neutron does not, but the energy in a proton's electric field is much smaller than the energies holding the proton and neutron together in the first place. So the model developed at that time, called the isospin model, said that protons and neutrons are in some way two different forms of what is fundamentally the same particle, the nucleon. This model, while obviously an approximation, is good enough to still be in widespread use today.

We begin by representing a nucleon when it is a proton as $\begin{bmatrix} 1 \\ 0 \end{bmatrix}$ and when it is a neutron as $\begin{bmatrix} 0 \\ 1 \end{bmatrix}$. In general, a nucleon N will be

$$N = a \begin{bmatrix} 1 \\ 0 \end{bmatrix} + b \begin{bmatrix} 0 \\ 1 \end{bmatrix}. \tag{55}$$

The numbers a and b are complex numbers. They are the quantum mechanical amplitudes for the nucleon N to be either

127

a proton or a neutron. The model is that while the proton *vs.* neutron nature of a nucleon might be known when it interacts with another particle, in between those interactions there are two interfering alternatives. Of course, if we allow that the neutron and the proton have different charges, then interference is not possible, because the alternatives can indeed be distinguished. But the fundamental assumption of this model is to neglect the electrical charges altogether.

Notice how we use the word model here, rather than the word theory. Both words refer to a mental construction that is designed to describe reality, but the usage is different. Scientists use the word theory for a mental construction that is very well established; it has to have solid experimental data to support it, and it has to have been studied in detail and found not to have any weak points. A model lacks that certainty. In this case, we are ignoring the small difference in the masses, and the difference in electrical charges between the proton and the neutron. Isospin is the same sort of thing as a theory, but it is not really good enough to be called a theory; it is just a model.

We will need to have a function of a nucleon that will reveal what the proton component is, and, similarly, we need a function to say what the neutron component is. These functions input a column vector and output a complex number. We will call them projection functions and write them as f_p and f_n. Starting with some nucleon constructed with, say, $a = 0.6$ and $b = 0.8i$, then

$$f_p\left(\begin{bmatrix} 0.6 \\ 0.8i \end{bmatrix}\right) = 0.6$$

and

$$f_n\left(\begin{bmatrix} 0.6 \\ 0.8i \end{bmatrix}\right) = 0.8i. \tag{56}$$

For this particular nucleon, the amplitude for it to appear as a proton when it is observed is 0.6, and the amplitude for it to appear as a neutron when it is observed is $0.8i$. The proton component of the proton is 1, the neutron component of the neutron is 1, the proton component of the neutron is 0 and the neutron component of the proton is also 0.

As a function, f_p takes its input vector and does a matrix multiplication with the row matrix $[1\ 0]$, and f_n similarly is a function that takes its input and does matrix multiplication with $[0\ 1]$. It is common to gloss over the fact that these are functions and to just say express them in terms of dot products. However, projection functions are not really vectors; they can have different units than the vectors they operate on, and do not always transform with rotations the way true vectors do, according to Equation (28). Accordingly, it is also a common practice to call true vectors "contravariant vectors" and to call the projection functions "covariant vectors". Projection functions are also called 1-forms or covectors.

Now that we have f_p and f_n as projections of the proton and neutron components, should not there be an f_N, a projection of the component of an arbitrary nucleon? Indeed there is such a projection function. The definition that works is

$$f_N(N) = a^*a + b^*b, \quad \text{when } N = \begin{bmatrix} a \\ b \end{bmatrix}, \tag{57}$$

and for the N component of N to be 1, it must be that $a^*a + b*b = |a|^2 + |b|^2 = 1$.

Chapter 32

Your First Nuclear Physics Theory: Symmetry

You might have asked, why use two-row column vectors to represent nucleons? Would not the numbers 1 and 0 do? Or some other pair of distinct symbols? Should we perhaps label the proton and nucleon with the elements of the group Z_2? The reason for this choice is to make it easy to mathematically implement certain symmetry properties. These nucleons N are analogous to the square that we used when introducing groups; they are the object being studied. Then we ask, "What are the transformations that leave the nucleon unchanged?"

An obvious set of transformations of column vectors with two rows are the 2×2 matrices that can be applied to them. After all, a matrix can be used as a function of a column vector that outputs a column vector. Using U to be the 2×2 matrix that is the transform, we are looking at replacing nucleons N with transformed nucleons UN to implement our symmetry.

The matrices that will express some kind of symmetry are the ones that can be applied without changing anything physically meaningful. All the physical values that can be calculated before the transform is applied must be the same after the transformation. That means that the proton component of an proton must not be changed by applying the matrix U to the vector for a proton and the neutron component of a neutron must not be changed by applying the matrix U to a neutron. Generally, the value of the f_N

projection function must not change when the matrix U is applied to an arbitrary nucleon N.

The projection functions also need to change. When N in Equation (55) is replaced by a transformed nucleon UN,

$$UN = \begin{pmatrix} u_{11} & u_{12} \\ u_{21} & u_{22} \end{pmatrix} \begin{bmatrix} a \\ b \end{bmatrix} = \begin{bmatrix} u_{11}a + u_{12}b \\ u_{21}a + u_{22}b \end{bmatrix} \tag{58}$$

Equation (57) changes also, to

$$f_{UN}(UN) = (u_{11}a + u_{12}b)^*(u_{11}a + u_{12}b) + (u_{21}a + u_{22}b)^*(u_{21}a + u_{22}b). \tag{59}$$

In order for the projection function f_{UN} when applied to UN, that is, $f_{UN}(UN)$, to be the same as the projection function f_N when applied to N, that is, $f_N(N)$, it must be that

$$a^*a + b^*b = (u_{11}a + u_{12}b)^*(u_{11}a + u_{12}b)$$
$$+ (u_{21}a + u_{22}b)^*(u_{21}a + u_{22}b) \tag{60}$$

which will only be true for arbitrary a and b if

$$\begin{aligned} u_{11}^*u_{11} + u_{21}^*u_{21} = 1 \quad u_{11}^*u_{12} + u_{21}^*u_{22} = 0 \\ u_{12}^*u_{11} + u_{22}^*u_{21} = 0 \quad u_{12}^*u_{12} + u_{22}^*u_{22} = 1 \end{aligned} \tag{61}$$

Exercise 29. Get from Equation (60) to Equations (61).

The succinct way to write Equations (61) is

$$U^\dagger U = \begin{bmatrix} 1 & 0 \\ 0 & 1 \end{bmatrix}, \tag{62}$$

where $U^\dagger$ is the Hermitian conjugate of U. To construct $U^\dagger$, take the complex conjugate of every element of U and then transpose the result.

Exercise 30. Show that for $U = \begin{pmatrix} u_{11} & u_{12} \\ u_{21} & u_{22} \end{pmatrix}$,

$$U^\dagger U = \begin{pmatrix} u_{11}^*u_{11} + u_{21}^*u_{21} & u_{11}^*u_{12} + u_{21}^*u_{22} \\ u_{12}^*u_{11} + u_{22}^*u_{21} & u_{12}^*u_{12} + u_{22}^*u_{22} \end{pmatrix}.$$

A matrix where $U^\dagger U$ is the unit matrix, $\mathbb{1}$, is called a unitary matrix.

Here is the symmetry then, in terms of our three formal steps:

(1) *Say something describing what we observe*: In this case, the initial statement is that for some nucleon, N, its own projection function $f_N(N)$ has some specific value, which happens to be 1.
(2) *Transform what we are looking at in some particular way*: The transformation is to multiply N by a unitary 2×2 matrix U, and to change the projection function to f_{UN} correspondingly.
(3) *Ask if we can still say the same thing we said in step 1. If so, we have symmetry*: What we have worked out is that if U is unitary, then indeed, $f_{UN}(UN) = f_N(N)$ is still that specific value of 1.

As it happens, the matrices U will form a group in and of themselves; but the real point is that these matrices provide the symmetry wherein the behavior of the projection functions is unchanged.

Chapter 33

$SU(2)$: A Matrix Group

The isospin symmetry therefore is that our nucleons N are unchanged when multiplied by U, (and their projection functions change correspondingly) where U is a unitary matrix. Actually, we are interested in the special subset of all such unitary matrices U that have a determinant of 1. That subset is the group called $SU(2)$; the S stands for special, to say that the determinant is 1.

A group has the four properties of closure, associativity, the existence of an identity and the existence of an inverse for every element in the group. Earlier, in Rotational Symmetries and Matrices, we showed that if a set of square matrices of some specific size has nonzero determinant, then the requirements of associativity and the existence of identity and inverses is satisfied. We just need to show closure.

If we have one matrix, A that is unitary, i.e., $A^\dagger A = \mathbb{1}$, and another matrix, B, which is also unitary, will the product AB also be unitary? The answer is yes:

$$(AB)^\dagger(AB) = B^\dagger A^\dagger AB = B^\dagger \mathbb{1} B = B^\dagger B = \mathbb{1}. \tag{63}$$

Voilà. We have a group.

Now to find the specific form for these matrices. Start with a 2×2 complex matrix, $\begin{bmatrix} a & b \\ c & d \end{bmatrix}$, where a, b, c and d are complex numbers. If the Hermitian conjugate and the inverse are equal, then, using

Equation (32),

$$\begin{bmatrix} a^* & c^* \\ b^* & d^* \end{bmatrix} = \frac{1}{ad-bc}\begin{bmatrix} d & -b \\ -c & a \end{bmatrix}.$$ (64)

Because the determinant $ad - cb = 1$,

$$\begin{bmatrix} a^* & c^* \\ b^* & d^* \end{bmatrix} = \begin{bmatrix} d & -b \\ -c & a \end{bmatrix}.$$ (65)

By matching the individual elements of the left side of Equation (65) to the corresponding elements on the right side, reveals that $a^* = d$ and $c^* = -b$. Therefore our group $SU(2)$ is the set of the matrices of the form

$$\begin{bmatrix} a & b \\ (-b)^* & a^* \end{bmatrix}$$ (66)

and the determinant $|a^2| + |b^2| = 1$.

Every element of $SU(2)$ can be written as a combination of its three generators. The only parameterization that ever gets used is

$$s_1 = i\sigma_1 = \begin{bmatrix} 0 & i \\ i & 0 \end{bmatrix} \quad s_2 = i\sigma_2 = \begin{bmatrix} 0 & 1 \\ -1 & 0 \end{bmatrix} \quad s_3 = i\sigma_3 = \begin{bmatrix} i & 0 \\ 0 & -i \end{bmatrix}.$$ (67)

where the σ_i are the three Pauli matrices,

$$\sigma_1 = \begin{bmatrix} 0 & 1 \\ 1 & 0 \end{bmatrix} \quad \sigma_2 = \begin{bmatrix} 0 & -i \\ i & 0 \end{bmatrix} \quad \sigma_3 = \begin{bmatrix} 1 & 0 \\ 0 & -1 \end{bmatrix}.$$ (68)

Exercise 31. Show that the Pauli matrices are not elements of the group $SU(2)$.

Certain combinations of the Pauli matrices have particular significance. One combination, $\sigma^+ = (\sigma_1 + i\sigma_2)/2$, will take a nucleon that is a neutron and make it into a proton:

$$\sigma^+ n = \begin{bmatrix} 0 & 1 \\ 0 & 0 \end{bmatrix} \begin{bmatrix} 0 \\ 1 \end{bmatrix} = \begin{bmatrix} 1 \\ 0 \end{bmatrix}. \tag{69}$$

Exercise 32. Define $\sigma^- = (\sigma_1 - i\sigma_2)/2$. Apply it to a proton.

Now for an interesting aside. Equation (55) is not only a way to represent a nucleon, but it is also the mathematical definition of a qubit, a bit as used in quantum computing.

In ordinary digital computing, numbers are represented in binary. A bit, mathematically speaking, is either a 0 or a 1. Logical gates are operations on the bits; for example, a NOT gate will turn a 0 into a 1 and vice-versa. Physically speaking, binary bits are nearly always implemented as a voltage at a certain point in an electrical circuit.

A bit in quantum computing is a device that can be represented by Equation (55). It is defined with two complex numbers, a and b. Actually, because phase invariance means that the phase of a quantum system is arbitrary, the phase of a can be set to zero, $\Im(a) = 0$ and only b would have a phase. Then the constraint $|a|^2 + |b|^2 = 1$ becomes $\Re(a)^2 + \Re(b)^2 + \Im(b)^2 = 1$. This constraint can then be used to determine one of these three remaining numbers; say, $\Re(a)^2$. So a qubit actually contains, with some sign ambiguities, two real numbers.

Single bit quantum gates are represented by 2×2 matrices, such as the Pauli matrices. The σ_3 matrix for example, is known to quantum computers as an R_π gate, sometimes also called a Z gate. The σ_1 matrix is the quantum version of a NOT gate.

Physically, a qubit is much harder to build than a binary bit. Nucleons do not make practical qubits using isospin. It is possible to use the magnetic field of a nucleon to build a qubit. People are also making, or trying to make, qubits out of superconductors, optical interferometers, ions trapped in space with electric and magnetic fields, defects in diamond created by substituting a nitrogen atom for a carbon atom, and quantum-dot-like structures built in silicon that have room for just a single electron. None of these methods work very well though! These tiny particles quickly lose the information that they are supposed to represent. As a result, error correction is a major topic in quantum computing.

Chapter 34

Your First Nuclear Physics Theory: Pions

What does it mean to apply σ^+ to a nucleon? This is what Hideki Yukawa pondered in 1935; he suggested that if you did apply σ^+ to a nucleon then you were describing a particle that would turn a neutron into a proton if it collided with the neutron. That particle would have to be positively charged, of course. Furthermore, σ^- describes a negatively charged particle that would turn a proton into a neutron if it collided with the proton.

For certain reasons, Yukawa[1] concluded that the mass of these charged particles would be somewhere between 2 and 20% of the mass of a proton. In 1936, a charged particle with a mass of about 11% of the proton's was found and named the muon. Thing is though, the muon does not interact with protons and neutrons in the way that Yukawa predicted. So the muon was an odd sort of a particle at that point in time; nobody really knew what to make of it. Isidor Rabi likened it to a dish that appeared at your table mysteriously, saying "Who ordered that?"

Yukawa thought about σ^+ and σ^- and postulated the existence of two charged particles. The fact that the Pauli matrices are basically two of the three generators for the $SU(2)$ group was perhaps known, but it was not widely appreciated at that time. If it were, Yukawa might have thought about σ_3, and might have correctly predicted

---

[1]H. Yukawa, "On the Interaction of Elementary Particles. I.", *Proc. Phys. Math Soc. Japan*, 17, 48 (1935).

three, rather than two, particles that interact with nucleons; a positively charged one, a negatively charged one, and a neutral one. This trio of particles does actually exist and was found in 1947. They are called pions, π^+, π^-, and π^0. They are about 14% of the mass of the proton.

Because they are less massive than the neutron and proton, particles such as the pion and others were collectively dubbed "mesons", from an ancient Greek word for "middle". The neutron and protons were classified as "baryons", from an ancient Greek word for "heavy". Because the electron is much lighter than the mesons, it is a "lepton", from an ancient Greek word for — you guessed it — "light".

In the isospin theory, there is another name for the pions. They are called "mediators". This name reflects the fact that they carry units of isospin from one nucleon to another. This term reflects really their role in this model.

There is a little complication. The complication is that the summed mass of a charged pion and a neutron is 114% of the mass of the proton that that is produced by the pion and neutron together. Therefore, the reaction which we want to describe with Equation (69), while it does indeed produce a proton, also produces energy in amount of 14% of a proton mass times c^2. That energy can come out as additional particles which have no isospin (such as gamma rays) or perhaps in some other form. Now that we've explained this, we will just gloss it over. Whenever we talk about how this particle collides with that particle to make something, there will be some other energy of some form produced as a side effect that we do not care about. We will not try to keep track of conservation of energy here, just to make it simpler to think about how isospin works.

As a result of work done in the 1960s and 1970s, which we will describe later, physicists now have a different understanding of protons, neutrons and pions, compared to Yukawa's time. Protons, neutrons and pions are composed of quarks, antiquarks and the energy that holds them together. The quark model is constructed so as to reproduce the symmetry of the isospin model. Quarks have an isospin symmetry of their own, which relates to the weak force; it is called weak isospin. The isospin symmetry of neutrons and protons relates to the strong force but is not usually called strong isospin. It is just called isospin, because that is the name it got in the years before the weak force was understood.

Chapter 35

Your First Particle Physics Theory: The Λ

By the early 1960s, it became apparent that there was some other quantity besides electric charge that some elementary particles had. It was not clear what this property was, exactly; particles that had it did not decay for a long time, and for lack of any better name this property was called strangeness. For example, there is a particle, called the lambda, Λ, which is about the same mass as a nucleon but is also strange. Actually, the convention nowadays is that the lambda has −1, rather than +1, units of strangeness. It also has no units of electrical charge.

It is convenient now to introduce a new unit of energy and mass. Imagine two points in space, and an electric field is such that the points differ in electric potential by a Volt. These might be the two terminals of a 1 V battery, for example. The energy of moving a Coulomb of charge between these points is a Joule, by definition. The energy of moving an electron, which has charge 1.60218×10^{-19} C, between those two points is 1.60218×10^{-19} J. This unit of energy is called the electron-Volt, eV. A million of them is an MeV. The energy in an MeV corresponds to a mass as well; that mass is MeV/c^2. The MeV is a small amount of energy by human standards; a standard AAA battery contains about 3×10^{16} MeV of energy.

We said that the Λ has "about the same mass" as the proton and the neutron. The proton's mass is 938.3 MeV/c^2, and the neutron's is 939.6 MeV/c^2. Those two are quite close. The Λ weighs in at 1115.7 MeV/c^2, about 19% more massive than a proton, and there

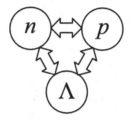

Figure 35.1: The extension of $SU(2)$ to $SU(3)$, part 1: The Λ interloper.

is an interesting tale that hangs from this. We will have to defer the tale for a little while, until Another Symmetry That Is Not.

Figure 35.1 expresses the Sakata model.[1] We already know what it means to go from the neutron to the proton, from the upper left corner of the triangle to the upper right corner. Physically, it means we collide a π^+ into a n, getting a p : $\pi^+ + n \to p$. Mathematically, it means we apply σ^+ to $\begin{bmatrix} 0 \\ 1 \end{bmatrix}$. Going from the proton to the neutron is similar: we apply σ^- to $\begin{bmatrix} 1 \\ 0 \end{bmatrix}$, getting $\begin{bmatrix} 0 & 0 \\ 1 & 0 \end{bmatrix} \begin{bmatrix} 1 \\ 0 \end{bmatrix} = \begin{bmatrix} 0 \\ 1 \end{bmatrix}$ But now there is a third possible state.

$$p = \begin{bmatrix} 1 \\ 0 \\ 0 \end{bmatrix}, \quad n = \begin{bmatrix} 0 \\ 1 \\ 0 \end{bmatrix} \quad \Lambda = \begin{bmatrix} 0 \\ 0 \\ 1 \end{bmatrix}. \tag{70}$$

We ask what does it mean to go from n to Λ or from Λ to p? Clearly, it involves changing the strangeness number; in the case of going from Λ to p, electrical charge is changed too.

The symmetry group $SU(2)$ must be extended to $SU(3)$, the set of all 3×3 unitary matrices with determinant 1. By this time, the importance of groups was better understood, and so it took little time to realize that because $SU(3)$ has eight generators, there must be eight particles that carry the force between the p, n and Λ. Just as three generators of $SU(2)$ are proportional to the three Pauli matrices, the eight generators of $SU(3)$ are proportional to the eight

---

[1] "On a Composite Model for the New Particles", Progress of Theoretical Phys., 16, 686 (Dec 1956).

Gell-Mann matrices, which are extensions of the Pauli matrices. The first three Gell-Mann matrices are

$$\lambda_1 = \begin{bmatrix} 0 & 1 & 0 \\ 1 & 0 & 0 \\ 0 & 0 & 0 \end{bmatrix}, \quad \lambda_2 = \begin{bmatrix} 0 & -i & 0 \\ i & 0 & 0 \\ 0 & 0 & 0 \end{bmatrix} \quad \lambda_3 = \begin{bmatrix} 1 & 0 & 0 \\ 0 & -1 & 0 \\ 0 & 0 & 0 \end{bmatrix}.$$

(71)

which basically are the three Pauli matrices, extended to have a third row and third column that do nothing.

Now it obvious to construct $\lambda^+ = (\lambda_1 + i\lambda_2)/2$ and say, ah yes! Just like $\sigma^+ = (\sigma_1 + i\sigma_2)/2$; this matrix must correspond to the π^+ meson for this extended theory. Similarly, $\lambda^- = (\lambda_1 - i\lambda_2)/2$ must correspond to the π^-, and λ_3 the π^0.

Exercise 33. What is λ^+ applied to n as given by Equation (70)? What is λ^- applied to p?

Your First Particle Physics Theory: Strange Mesons

What about the other five Gell-Mann matrices? Do not they have to correspond to five other mesons, rather like the $\pi^{\pm}$ and π^0? Indeed they do, and indeed those particles exist. These particles are the mediators of the extension of the isospin model to include strangeness. There are four particles, known as kaons, K^+, K^-, K^0, and $\bar{K}^0$, that are very much like the three pions, but they carry units of strangeness. This set of seven mesons can be laid out as shown in Figure 36.1. Particles on the upper right diagonal edge (K^+, π^+) are positively charged and so increase isospin; particles on the lower left diagonal edge (π^-, K^-) decrease isospin. Extending the idea that a π^+ and a n will make a p, a positive kaon (K^+) and a Λ should also form a p; $K^+ + \Lambda \to p$. Particles on the top edge (K^0, K^+), carry $+1$ units of strangeness, and on the bottom edge $(\bar{K}^0, K^-)$, -1 units of strangeness. So we expect that, for example, a $\bar{K}^0 + n \to \Lambda$. The particle in the center, the π^0, carries neither electrical charge nor strangeness.

We are still missing a meson. There are eight generators, but Figure 36.1 gives us only seven particles. There should be one more meson. There are two known particles that make good candidates; the η, and the η'. They are both uncharged and have no strangeness.

We will have to get to the quark structure of mesons before we can give the explanation for the η and η' story. The model that we are

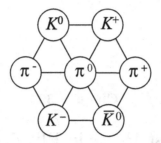

Figure 36.1: The extension of $SU(2)$ to $SU(3)$, part 2: A honeycomb of mesons.

using, where mesons are fundamental particles that change the state of the baryons from n to p to Λ, does not have space for a ninth meson. It is possible to explain the η' after learning that mesons and baryons are compound particles and that the constituents of the mesons themselves can be described with $SU(3)$.

Chapter 37

The Eightfold Way and Quarks

In the late 1950s and early 1960s, new particles were being discovered at a startling rate, and it was a real challenge to make sense of them. Gell-Mann, Ne'eman and Zweig were eventually able to make sense of at least the mesons and baryons with a model that Gell-Mann dubbed the "Eightfold Way". Let's start by looking at a certain specific group of baryons that are heavier than the nucleons and the Λ. Their properties are listed in Table 37.1. The lightest of these baryons is the Δ, so we will call these the "Δ-like" baryons; the three baryons which we know already (p, n and Λ) we will call the light baryons.

There are four Δ-like baryons with no strangeness, three with -1 units of strangeness, two with -2 units of strangeness and one with -3 units of strangeness. Every time the amount of strangeness changes, the mass changes. It is as if there were strange (or antistrange!) things inside these baryons that were a little heavier than the other things inside them and that the total number of strange things that you can add is three. By comparing the masses of particles with different strangeness values, you can see that the mass of this strange internal component of these particles is about $150 \, \mathrm{MeV}/c^2$ more than the mass of whatever it replaces.

Look only at the particles that have no strangeness, the Δ particles, there are four possible values of charge, just as there are four possible values of strangeness. There is a negative Δ, a neutral Δ, a positive Δ and even a Δ with a double-positive

149

Table 37.1: Δ-like baryons. The average of the Σ^{*+} and Σ^{*-} masses is given.

Name	Charge	Strangeness	Mass (MeV/c)^2
Δ^0	0	0	$\sim$1232
$\Delta^\pm$	± 1	0	$\sim$1232
Δ^{++}	2	0	$\sim$1232
$\Sigma^{*\pm}$	± 1	-1	1385.0
Σ^{*0}	0	-1	1383.7
Ξ^{*0}	0	-2	1531.8
Ξ^{*-}	-1	-2	1535.0
Ω^-	-1	-3	1672.5

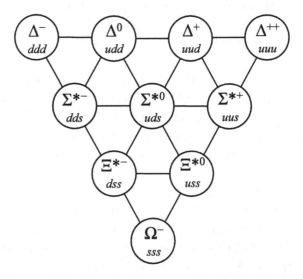

Figure 37.1: The quark assignment for Δ-like baryons.

charge — as though there is a thing that you can put into these particles that increases its charge but not its strangeness or mass.

These things that you put into these particles were called aces by Zweig and quarks by Gell-Mann. There will be three of them in each Δ-like baryon. They are named up, u, down, d, and strange, s. Figure 37.1 shows the assignment of quarks in each of these particles.

The last of the Δ-like baryons is well named. At the 1962 International Conference on High Energy Physics — in those days it was still often called the Rochester conference — at CERN, evidence was presented for the Ξ^* hyperon from the Brookhaven and Lawrence Berkeley labs.[1] Gell-Mann, during the conference, predicted the existence, mass, decay modes, spin, charge, and strangeness of the Ω^-, along with the fact that it would be the last of the Δ-like baryons, as all the others had been found by then. The Ω^- was discovered in 1964 at Brookhaven.[2]

Exercise 34. From the quark composition of the Δ^{++}, Δ^- and Ω^-, find the charges of the u, d and s quarks.

Exercise 35. From the quark composition of the Δ^{++}, Δ^- and Ω^-, find the strangeness values of the u, d and s quarks.

The first key idea of the Eightfold Way is that these three quarks have an $SU(3)$ symmetry of their own.

$$u = \begin{bmatrix} 1 \\ 0 \\ 0 \end{bmatrix}, \quad d = \begin{bmatrix} 0 \\ 1 \\ 0 \end{bmatrix} \quad s = \begin{bmatrix} 0 \\ 0 \\ 1 \end{bmatrix}. \tag{72}$$

The mathematics here is the same as what we developed to model the p, n, Λ triad, but now we are applying it to a different set of three objects. Now we are saying that there are these three quarks and they have $SU(3)$ symmetry; earlier we were saying that there are these three baryons and they have $SU(3)$ symmetry. As before, the symmetry is not exact; the Λ is heavier than the p and n, and the s is heavier than the u and d.

---

[1]L. Bertanza, V. Brisson, P. L. Connolly, E. L. Hart, I. S. Mittra, G. C. Moneti, R. R. Rau, N. P. Samios, *et al.* "The K$^-$p interaction at 2.24 GeV/c I: Effective mass distributions" in Proceedings of the 11th International Conference on High-energy Physics (Geneva, Switzerland, 4–11 July 1962), http://inspirehep. net/record/1341794/; G. M. Pjerrou, D. J. Prowse, P. Schlein, W. E. Slater, D. H. Stork, H. K. Ticho, "A resonance in the $\Xi\pi$ system at 1.53 GeV", http://inspirehep.net/record/1341796/; Discussion of G. Snow, "Strong interactions of strange particles", p. 805, http://inspirehep.net/record/1341918/.
[2]V. E. Barnes, P. L. Connolly, D. J. Crennell, B. B. Culwick, W. C. Delaney, W. B. Fowler, P. E. Hagerty, E. L. Hart, *et al.* "Observation of a Hyperon with Strangeness Minus Three", *Phys. Rev. Lett.*, 12, 204 (1964).

By analogy with the $SU(3)$ baryon symmetry with its eight mesons, physicists expected that there would be eight particles that might, for example, strike a u quark and turn it into a d or s quark. There should be eight mediators for the quark symmetry. There are indeed such mediating particles, but there are not eight of them. It turns out that while the Eightfold Way was a great step forward, and explains the existence of quite a few mesons and baryons, it is not the final answer either, and that is why these eight particles do not exist.

The quark composition of the light baryons, n, p, and Λ, is similar to that of Δ^0, Δ^+, and the Σ^{*0}, respectively. The neutron n is udd, the proton p is uud and Λ is uds. The arrangement of the quarks inside the light baryons is different from the arrangement of the quarks in the Δ-like baryons. Baryons are fermions; recall from About Spin that fermions have spin $n\hbar/2$, where n is an odd number. The quarks inside the light baryons are arranged to make a total overall spin of $\hbar/2$; for the Δ-like baryons, the overall spin is $3\hbar/2$.

The second idea of the Eightfold Way is that quarks have antimatter versions. The key for us is that mesons are made of one quark and one antiquark. The quark-antiquark components of the mesons are shown in Figure 37.2.

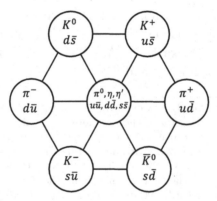

Figure 37.2: Quark-antiquark assignments for the mesons of Figure 36.1, the η and the η'.

The center of Figure 36.1 shows three mesons, the π^0, the η, and the η', which are mixtures of $u\bar{u}$, $d\bar{d}$ and $s\bar{s}$. Three types of quarks, three types of quark-antiquark pairs, three mesons. It takes quarks to explain the existence of the η'!

It takes more than just quarks to explain the masses of the π^0, the η, and the η', however. The π^0 has a mass of just $135.0 \, \text{MeV}/c^2$, while the η has a mass of $547.9 \, \text{MeV}/c^2$ and the η', has a mass of $957.8 \, \text{MeV}/c^2$. The intrinsic masses of the u and d quarks are actually quite small; they are on the order of $10 \, \text{MeV}/c^2$ or less. The s quark's mass is around $100 \, \text{MeV}/c^2$. What is happening is that the forces that bind the quarks together are quite strong — indeed, this is the strong force that we first discussed in A Symmetry That Is Not. Relatively large amounts of energy are involved, and this energy forms the bulk of the masses of all the mesons and baryons that can be made from u, d, and s quarks and antiquarks. In fact, there is so much mass associated with the forces that it is difficult to determine what the bare masses of the quarks themselves are. Consequently, the masses of not only these three mesons, but of all the mesons and baryons composed of u, d, and s quarks depends primarily upon how the quarks are arranged in the mesons and baryons, and not on the masses of the quarks themselves.

Now we are able to understand the reactions of Yukawa's isospin model and its extension to include the Λ in terms of the quarks and antiquarks inside of the baryons and mesons. For example, writing the reaction $\bar{K}^0 + n \to \Lambda$ into its quark components, $s\bar{d} + udd \to uds$; a d quark and the $\bar{d}$ antiquark annihilate each other. The reaction $\pi^+ + n \to p$ is $u\bar{d} + udd \to uud$, and $K^+ + \Lambda \to p$ is $u\bar{s} + uds \to uud$.

Let us recap our $SU(n)$ symmetries so far.

(1) We used $SU(2)$ for the proton and neutron and found that the three generators meant that there were three π mesons.

(2) We used $SU(3)$ for the p, n, and Λ and found that the eight generators of $SU(3)$ meant that there were eight mesons, including the three pions.

(3) We used $SU(3)$ for the u, d, and s quarks and failed to find eight particles corresponding to the eight generators. Those particles should be there if these quarks really have $SU(3)$ symmetry, so we had to discard that theory. However, we were able to categorize a large number of particles in terms of their quark content.

Chapter 38

Another Symmetry that Is Not

There were some problems with the quark theory that delayed its widespread acceptance from the early 1960s until 1974. One of the issues was that there were number of searches for free quarks, outside of mesons and baryons, and they all failed. On the other hand, starting in 1967, experiments at SLAC started to see that the interior of proton was not uniform; there seemed to be some kind of hard objects inside the proton. The quark theory not being widely accepted at that time, these were not called quarks. Instead, as they were known only to be parts of the proton they were called, in a not very imaginative terminology, partons.

In 1973, D. Gross, F. Wilczek and H. Politzer[1] solved certain issues with the theory of the forces between quarks based on $SU(3)$ symmetry and color charges, making the idea plausible. This theory was based on the idea that there were three charges for the strong force — red, green and blue, as described in Antimatter — and that there is an $SU(3)$ symmetry for these three colors. The eight mediating particles for this symmetry are the gluons. This theory solved a couple of problems. Most notably, it explained why there were no free quarks. It also explained why there are three-quark combinations and quark-antiquark combinations but not, for example, two-quark or four-antiquark combinations. By November 1974, there were not

---

[1] D. Gross and F. Wilczek, "Ultraviolet Behavior of Non-Abelian Gauge Theories", *Phys. Rev. Lett.*, 30, 1343 (1973); H. D. Politzer, "Reliable Perturbative Results for Strong Interactions", *Phys. Rev. Lett.*, 30, 1346 (1973).

any real problems with the quark theory. However, that is not the same as saying it was proven, or even widely accepted.

Then a new particle was found.[2] It was seen at Brookhaven by a group led by Sam Ting and at SLAC by a group led by Burt Richter. Within about a year, this particle was understood as being composed of a new, fourth quark, and its antiparticle.

Immediately, physicists thought "$SU(4)$! For four quarks!". Not for very long though. For one thing, there was the mass asymmetry. Remember that the nucleons of the $SU(2)$ isospin theory, p and n, are nearly equal in mass. The strange baryons — that is, the Λ, Σ^*, Ξ^*, and Ω^- baryons — are not really equal in mass to the p and n, but they are close enough that $SU(3)$ can work with a little allowance for the mass differences. But the fourth quark, the charm quark, c, is very different in mass; a single c has a mass of $1270\,\mathrm{MeV}/c^2$, much more than the u, d or s quarks. The charm quark by itself is heavier than many baryons that contain three light quarks. For another thing, there should have been eight mediating particles that exist as a result of $SU(3)$ quark symmetry having eight generators, but which do not exist. Trying $SU(4)$ for quarks makes that problem worse! That is because $SU(4)$ has 15 generators, and so now there are seven more missing mediating particles. It is possible to apply $SU(4)$ and figure out what mesons and baryons should (and do) exist, as a result of mesons being made of quark-antiquark pairs and baryons being made of three quarks. But really, the whole idea that $SU(n)$ symmetry applies to quarks as a fundamental law of physics really doesn't work.

In 1977, a fifth quark was discovered at Fermilab in an experiment led by Leon Lederman;[3] it was even more massive, at $4180\,\mathrm{MeV}/c^2$.

---

[2]J. J. Aubert, U. Becker, P. J. Biggs, J. Burger, M. Chen, G. Everhart, P. Goldhagen, J. Leong, *et al.*, *Phys. Rev. Lett.*, 33, 1404 (1974); J. E. Augustin, A. M. Boyarski, M. Breidenbach, F. Bulos, J. T. Dakin, G. J. Feldman, G. E. Fischer, D. Fryberger, *et al.*, *Phys. Rev. Lett.*, 33, 1406 (1974).

[3]S. W. Herb, D. C. Hom, L. M. Lederman, J. C. Sens, H. D. Snyder, J. K. Yoh, J. A. Appel, B. C. Brown, *et al.*, *Phys. Rev. Lett.*, 39, 252.

And a sixth quark, the top quark, was found at Fermilab in 1995.[4] The top quark is *much* more massive than the three light quarks. The mass of the top is about $173,210 \, \text{MeV}/c^2$.

Even though $SU(n)$ for quarks does not really work, it did work fairly well for explaining the light and Δ-like baryons and their corresponding mesons. And $SU(3)$ works perfectly well for understanding the strong force between the quarks. This kind of symmetry was still clearly enough of a success to keep in mind.

---

[4]F. Abe, H. Akimoto, A. Akopian, M. G. Albrow, S. R. Amendolia, D. Amidei, J. Antos, C. Anway-Wiese, *et al.*, *Phys. Rev. Lett.*, 74, 2626 (1995); S. Abachi, B. Abbott, M. Abolins, B. S. Acharya, I. Adam, D. L. Adams, M. Adams, S. Ahn, *et al.*, *Phys. Rev. Lett.*, 74, 2632 (1995).

Chapter 39

γ, W, Z, and H

Now, what about $SU(1)$? That would be a 1×1 matrix, $[a]$, with the unitarity requirement $[a]^\dagger [a] = [1]$. That means $a^*a = 1$, so that a is a complex number of magnitude 1. That is our old friend, the circle group. It is called $U(1)$ rather than $SU(1)$ because the determinant of $[a]$ is not necessarily 1; it only has magnitude 1.

We found back in Gauge Invariance that the amplitudes of quantum mechanics have $U(1)$ symmetry. While trying to understand the quantum form of the electromagnetic field, physicists discovered that this symmetry was very central to the nature of electromagnetism. The phase ϕ in the element $e^{i\phi}$ of the circle group turns out to be a function of the electric and magnetic potentials and this factor of $e^{i\phi}$ appears in the amplitudes of quantum electromagnetism. Because $U(1)$ has only one generator there should be only one particle that takes the role of a mediator in electromagnetism. It is the photon.

Instead of a single six element group, $SU(6)$, for the six quarks, nature instead groups the six quarks into three pairs of $SU(2)$ symmetric particles. The pairs are (u, d), (c, s) and (b, t). Each pair is like the other two, except with different masses. The structure seems to be like the periodic table of the elements, except with only two columns and three rows.

Because the quarks form these pairs, and because there are three generators for their $SU(2)$ symmetry, there are three mediating particles that change for example u quarks into d quarks or t quarks into b quarks. These three particles, the mediators of the weak force, are the $W^{\pm}$ and the Z. They are very much like the photon, but are massive; they are about 100 times the mass of a proton.

Let us re-recap our $SU(n)$ symmetries, so far.

(1) We used $SU(2)$ for the proton and neutron and found that the three generators meant that there were three π mesons.

(2) We used $SU(3)$ for the p, n, and Λ and found that the eight generators of $SU(3)$ meant that there were eight mesons, including the three pions.

(3) We used $SU(3)$ for the u, d, and s quarks and failed to find eight particles corresponding to the eight generators. Those particles should be there if these quarks really have $SU(3)$ symmetry, so we had to discard that theory. However, we were able to categorize a large number of particles in terms of their quark content.

(4) We used $SU(3)$ for the red, green, and blue color charges of the strong force and found that the eight generators $SU(3)$ meant that there were eight mediating particles of the strong force, the eight gluons.

(5) We used $U(1)$ for the electromagnetic force, and its single generator corresponds to the single mediating particle of electromagnetism, the photon.

(6) We replaced our failed attempt to use $SU(3)$ for the u, d, and s quarks, and subsequently our even greater failure to use $SU(6)$ for all six quarks with three identical copies of $SU(2)$: (u, d), (c, s) and (b, t). The three mediators are the $W^{\pm}$ and the Z.

Now we can answer Rabi's question, "Who ordered that?" The muon, like the electron, is a lepton. Just as there are three pairs of quarks, there are three pairs of leptons. They are (ν_e, e^-), (ν_μ, μ^-), and (ν_τ, τ^-). The muon is a heavier version of the electron, just as the strange quark is a heavier form of the down quark. There is another lepton, found by a group lead by Martin Perl[1] in 1975, called the

---

[1]M. L. Perl, G. S. Abrams, A. M. Boyarski, M. Breidenback, D. D. Briggs, F. Bulos, W. Chinowskym, J. T. Dakin, *et al.*, "Evidence for Anomalous Lepton Production in e^+-e^- Annihilation", *Phys. Rev. Lett.*, 35, 1489 (1975).

τ^-; it is a still heavier version of the electron and muon. The three ν particles are neutrinos; they have neither electric nor color charges. Neutrinos are difficult particles to measure; physicists do not know a lot about them as a result.

There are two questions about the three weak force mediators. The first is... are there three sets of $W^\pm$ and Zs, one for each of the three pairs? The answer is no... there is only one set. Occasionally the interactions will jump between the three pairs. That is, occasionally a W^- will turn a t into a d rather than a b, or a W^+ will turn a d into a c rather than a u.

The second question about the three weak force mediators is a little less obvious. If a particle has inertia, if it has mass, then that affects its motion and the mass has to appear somehow in the Lagrangian. If you account for the masses of particles by putting the simple obvious thing into the Lagrangian, you destroy gauge symmetry. In 1964, Peter Higgs, Francois Englert, Robert Brout, Gerald Guralnik, Carl Hagen and Tom Kibble[2] worked out a mathematical method to keep gauge symmetry for massive $W^\pm$ and Z particles, and for the other massive fundamental particles as well. The trick is to suppose that there is a new field, like an electric or gravitational field, that is always present and is not zero, even in otherwise empty space. That field interacts with the other particles in such a way as to slow them down and keep them from moving at the speed of light.

The best explanation of this is from David Miller of University College, London:

Imagine a cocktail party of political party workers who are uniformly distributed across the floor, all talking to their nearest neighbors. The ex-Prime Minister enters and crosses the room. All of the workers in her neighborhood are strongly attracted to her

---

[2]P. W. Higgs, "Broken Symmetries and the Masses of Gauge Bosons", *Phys. Rev. Lett.*, 13, 508 (1964); F. Englert and R. Brout, "Broken Symmetry and the Mass of Gauge Vector Mesons", *Phys. Rev. Lett.*, 13, 321 (1964); G. S. Guralnik, C. R. Hagen and T. W. B. Kibble, "Global Conservation Laws and Massless Particles", *Phys. Rev. Lett.*, 13, 585 (1964).

and cluster round her. As she moves she attracts the people she comes close to, while the ones she has left return to their even spacing. Because of the knot of people always clustered around her she acquires a greater mass than normal, that is, she has more momentum for the same speed of movement across the room. Once moving she is harder to stop, and once stopped she is harder to get moving again because the clustering process has to be restarted. In three dimensions, and with the complications of relativity, this is the Higgs mechanism.

Higgs was the first to notice that if the mathematical trick was what actually happened in the real world, that there would be at least one new particle out there to be discovered. Those particles are the Higgs bosons;[3] one was found in 2012, although there could be, and physicists do keep looking for, more.

---

[3]The ATLAS Collaboration, "Observation of a New Particle in the Search for the Standard Model Higgs Boson with the ATLAS Detector at the LHC", *Phys. Lett.* B, 716, 1 (2012); CMS Collaboration, "Observation of a New Boson at a Mass of 125 GeV with the CMS experiment at the LHC", *Phys. Lett.* B, 716, 30 (2012).

Chapter 40

Bra-Kets

Earlier, we described protons and neutrons as column vectors with real entries — 1 and 0, to be precise. To describe a general nucleon, the entries were complex numbers, so the state of a nucleon can be described in this model with two complex numbers, as in Equation (55). We extended the idea to include strangeness, which is an example of $SU(3)$ symmetry. In that case, the state can be described with three complex numbers.

How many complex numbers are needed to specify a quantum state? It depends on both what physical properties you are interested in and how many possible outcomes there are. When a nucleon interacts with a pion, it might appear as a proton at the time of the interaction, or it might be a neutron at the time of the interaction. With two distinct, discrete outcomes, two numbers are needed. But to study the energy of a proton for example requires an infinite number of complex numbers, because the energy is described by a real number, and the real numbers are infinite.

Writing matrices is a little tedious, especially when they are infinite in size. A better notation would help. One of the best notations to handle this situation, and probably most problems in quantum mechanics, is the bra-ket notation of Dirac.

$$\langle bra \mid ket \rangle$$

The kets, the things written on the right side, correspond to the column vectors, which represent states. In our discussion of isospin and strangeness, all the states were specific particles. A nucleon N is written as $|N\rangle$; a proton or a neutron is written as a $|p\rangle$ or a $|n\rangle$. In ket notation, Equation (55) is replaced with

$$|N\rangle = a|p\rangle + b|n\rangle. \tag{73}$$

To speak of a particle of a specific energy E traveling through empty space, we could write a ket $|E\rangle$; you might think of this as a column vector of infinite size, with each element corresponding to a different specific energy. All of the elements but one would be zero; the element of the vector corresponding to that specific value of the energy would be one.

The bras of the bra-ket notation correspond to projection functions such as f_p, f_n and f_N. They are written $\langle p|$, $\langle n|$ and $\langle N|$, respectively. Instead of applying the function to a written-out column matrix, we write the bra for the function followed by the ket for the state. The proton projection of a nucleon N is $\langle p|N\rangle$. This is a single complex number — an amplitude.

In bra-ket notation, the application of f_p to a proton, $f_p(p) = 1$ is written as $\langle p|p\rangle = 1$. Because the application of f_p to a neutron yields no amplitude at all, $f_p(n) = \langle p|n\rangle = 0$. Consequently,

$$\langle p|N\rangle = \langle p|(a|p\rangle + b|n\rangle) = a\langle p|p\rangle + b\langle p|n\rangle = a. \tag{74}$$

If $a = 0.6$ and $b = 0.8i$, Equation (74) is the same as the first part of Equation (56).

In our two-slit experiment of The Quantum Mechanical Robert Frost, we found the amplitude for the electron to appear at a specific position x on the detector. That projection function is $\langle x|$. The state, the ket, is "The electron after it has passed through the two slits". Write it as $|2\,slits\rangle$. In fact, $|2\,slits\rangle = |Left\,slit\rangle + |Right\,slit\rangle$, and the fact that there are interfering amplitudes and that they sum is written as

$$\langle x|2\,slits\rangle = \langle x|Left\,slit\rangle + \langle x|Right\,slit\rangle. \tag{75}$$

Equation (75), in the notation of Chapters 7–10, is $A(x) = A_L(x) + A_R(x)$. To calculate a probability, take $\langle x|2\,slits\rangle$, which is a complex number, and multiply it by its complex conjugate.

In other words, a bra-ket combination with bra $\langle x|$ is the wavefunction; it is the sum of all the amplitudes that would result in an observation of the electron at x. If instead of projecting to the position x with the bra $\langle x|$, we project to the momentum p with the bra $\langle p|$, we have the amplitude for the particle to appear with momentum p.

Not only can you place a bra, a projection function, in front of a ket, you can also place an operator in front of a ket. An operator is a thing that inputs a ket, just like a bra, but outputs a ket rather than a complex number the way a bra does. The operator replaces the matrix in the same way that a ket replaces a column vector. The symbol for the operator is written to the left of the ket. In bra-ket notation, Equation (69) is written as

$$\sigma^+|n\rangle = |p\rangle. \tag{76}$$

Operators correspond to (perhaps infinitely large) matrices, and this means they must have the same arithmetic as matrices. It is possible to add operators, just as it is possible to add matrices. It is possible to multiply operators by numbers, just as it is possible to multiply a matrix by a number. And there is an operator multiplication, just as there is a matrix multiplication. For example, we must have $\sigma^+ = \sigma_1 + i\sigma_2$, whether σ^+, σ_1, and σ_2 are operators on kets or 2×2 matrices.

Operators can also form groups, with the usual properties of closure, associativity, and the existence of identity elements and inverses. The operators that correspond to the group of matrices $SU(n)$ will have an exact one-to-one correspondence to those matrices and the results of multiplying them. We will use $\mathbb{1}$ for the identity element for a group of operators, just as we did for matrices.

With matrices and column vectors, when we applied the operator U, which represented a hoped-for symmetry, to a nucleon N, the corresponding projection function f_N was transformed into f_{UN}. Similarly, whenever such an operator U is applied to the ket $|N\rangle$, then the projection function, the bras $\langle M|$, are transformed. We will write the transformed bras as $\langle M|U^\dagger$. Since operators are effectively matrices, and all matrices have Hermitian conjugates, we just go and write the operator $U^\dagger$ with impunity. It must exist, and the notation is designed to reflect Equation (62).

Now we work out $SU(2)$ symmetry, just as in Your First Nuclear Physics Theory: Symmetry", with the bra-ket notation.

(1) *Say something describing what we observe*: In this case, the initial statement is that the projection function of a nucleon $|N\rangle$ given by $\langle M|$ is $\langle M|N\rangle$.

(2) *Transform what we are looking at in some particular way*: The transformation is to apply the operator U to $|N\rangle$, obtaining $U|N\rangle$, and to change the projection function correspondingly. The projection function $\langle M|$ will be changed into a projection function written as $\langle M|U^\dagger$.

(3) *Ask if we can still say the same thing we said in step 1. If so, we have symmetry*: The symmetry appears if $\langle M|N\rangle = \langle M|U^\dagger U|N\rangle$, which will happen if and only if $U^\dagger U$ is the identity operator $\mathbb{1}$.

Step 3 tells us what $U^\dagger$ is. The operator U corresponds to some matrix; that matrix has a complex conjugate; that complex conjugate has a corresponding operator. The symmetry appears if the complex conjugate operator and the operator itself cancel, and $U^\dagger U = \mathbb{1}$.

Using matrices, we worked only out the $SU(2)$ case. With the bra-ket notation, we cover also the $SU(3)$ case, the $SU(4)$ case, and in fact any kind of transformation that represents any kind of symmetry. The only requirement is that the class of transformations operates on the states that we are interested in and is unitary. All that needs to be done is to write (and know!) that $\langle M|N\rangle = \langle M|U^\dagger U|N\rangle$.

Chapter 41

Two Fermions in a Pod

An important idea that exists in quantum mechanics that is not part of our normal everyday lives is the idea of identical indistinguishable particles. When there are two or more indistinguishable particles, there almost always are interfering, indistinguishable alternatives.

We might say "those two are peas in a pod", meaning that those two, whoever they might be, cannot be told apart. But peas are not subatomic particles. If you take any two peas out of their pod and examine them carefully, you will find some minute differences. Even if the differences are too small to see, you can at least identify the two peas by noting their position — you can speak of the pea over here and the pea over there.

With objects small enough to need quantum mechanical descriptions, this is no longer true. If you examine two protons or two electrons carefully, you will not find any minute differences whatsoever. Furthermore, you cannot really rely on identifying the particles by position; it can be that they are so close together that, according to Heisenberg's Uncertainty Principle, you cannot tell which one is over here and which one is over there.

Here is how to handle that situation quantum mechanically. Write the ket for particle 1 and particle 2 to be in some particular state as $|\psi_{12}\rangle$; then if the particles are swapped, call that the state $|\psi_{21}\rangle$. Ultimately, because these amplitudes are indistinguishable, it is their sum that is the ket for the complete two particle system.

Since the particles are perfectly identical, all the amplitudes that you can make out of one combination of the two particles you should be able to make out of the other combination of the two particles and get the same probabilities. Take any bra (we will use $\langle \vec{x} |$, but it does not matter) and construct the amplitude $\langle \vec{x} | \psi_{12} \rangle$. The probability $P_{12} \propto |\langle \vec{x} | \psi_{12} \rangle|^2$, and, again, because these two particles are indistinguishable from each other, P_{12} must be the same as the probability $P_{21} \propto |\langle \vec{x} | \psi_{21} \rangle|^2$. That is, swapping particle 1 with particle 2 should change none of the probabilities.

As a result of gauge invariance, these two amplitudes can differ only by a phase: $\langle \vec{x} | \psi_{21} \rangle = e^{i\delta} \langle \vec{x} | \psi_{12} \rangle$, or equivalently, $|\psi_{21} \rangle = e^{i\delta} |\psi_{12} \rangle$, where δ is some angle between 0 and 2π radians.

Exercise 36. Show that there can be only this phase difference.

If the particles are swapped two times, then you are back to the initial state, the phase factor δ from the swap introduced two times:

$$|\psi_{12} \rangle = e^{i\delta} |\psi_{21} \rangle = e^{i\delta} \left(|\psi_{21} \rangle \right) = e^{i\delta} \left(e^{i\delta} |\psi_{12} \rangle \right) = e^{2i\delta} |\psi_{12} \rangle. \qquad (77)$$

Therefore $e^{i2\delta} = 1$, implying that δ is either 0 or π. If $\delta = 0$, $|\psi_{12} \rangle = +|\psi_{21} \rangle$; if $\delta = \pi$, $|\psi_{12} \rangle = -|\psi_{21} \rangle$. Particles of the first type are called bosons; particles of the second type are called fermions.

Now consider the interfering alternatives produced by the fact that there are two indistinguishable cases. Because the particles are indistinguishable, the amplitude in any situation with two particles is the sum of the two alternatives:

$$\langle \vec{x} | \psi \rangle = \langle \vec{x} | \psi_{12} \rangle + \langle \vec{x} | \psi_{21} \rangle. \qquad (78)$$

For bosons, $|\psi_{12} \rangle = +|\psi_{21} \rangle$, and all is well and good. The sum of the two amplitudes is bigger by a factor of two, and the probabilities $|\langle \vec{x} | \psi \rangle|^2$ are bigger by a factor of four, but we said only that the probabilities are proportional to the amplitudes squared, so we can just change the constant of proportionality. But for fermions, $|\psi_{12} \rangle = -|\psi_{21} \rangle$, then $|\psi \rangle = 0$! If you try to put two identical fermions into the same state, all the amplitudes and therefore all the probabilities go to zero. You cannot do that! This is the Pauli Exclusion Principle; you cannot have two fermions in the same place at the same time.

The distinction between fermions and bosons is really fundamental. Particles that are little bundles of force fields, like photons, are bosons. In contrast, particles that respond to these force fields have antiparticles are fermions, or at least are composed of fermions. Quarks and leptons are fermions. The Higgs boson is (duh!) a boson. Pions and kaons in the $SU(2)$ and $SU(3)$ versions of isospin, carry the strong force, so they are bosons in those theories and, it turns out, in real life.

Chapter 42

The Back of the Book

Exercise 1. Suppose that there are two identity elements for a group. Call them I_1 and I_2, and then show that $I_1 = I_2$. Because I_1 is an identity, for any particular element A of the group, $I_1 \cdot A = A$. In particular, when A is I_2, we have $I_1 \cdot I_2 = I_2$. Similarly, $A \cdot I_2 = A$ and when A is I_1, we have $I_1 \cdot I_2 = I_1$. Since $I_1 \cdot I_2$ equals itself, $I_1 = I_2$.

Exercise 2. Suppose some element A in the group has two inverses, B and C. The problem is to show that $B = C$. From the definition of an inverse, $A \cdot B = \mathbb{1}$ and $A \cdot C = \mathbb{1}$ Then $A \cdot B = A \cdot C$. Multiply on the left on both sides by the inverse of A. That inverse times A is just the identity element $\mathbb{1}$. So now, $\mathbb{1} \cdot B = \mathbb{1} \cdot C$, which means that $B = C$.

Exercise 3. Suppose the contrary, that there are two different elements, B and C, in the row for element A that have the same entry in the table. That is, $AB = AC$. Multiply on the left by the inverse of A to get $A^{-1}AB = A^{-1}AC$. Then $B = C$; but the whole point is that B and C are different.

Exercise 4. Write an angle for each physical rotation; that is the set with which the group is defined. The binary operation on the group is to apply one rotation, followed by another. In terms of the angles, that operation is to add the number of radians in each angle. To show associativity, imagine three different rotations, R_1, R_2 and R_3. They are through angles θ_1, θ_2, and θ_3. Associativity is true if $(R_1 \cdot R_2) \cdot R_3 = R_1 \cdot (R_2 \cdot R_3)$. Equivalently, $(\theta_1 + \theta_2) + \theta_3 = \theta_1 + (\theta_2 + \theta_3)$ where we subtract or add a multiple of 2π radians until the result is between 0 and 2π. But this is true from the way addition works. In fact this is also an a Abelian group: $\theta_1 \cdot \theta_2 = \theta_2 \cdot \theta_1$, so $R_1 \cdot R_2$ is the same rotation as $R_2 \cdot R_1$. To show closure, realize that if you rotate a circle around its center, and then rotate it again, the net effect is still a rotation. To show the existence of an inverse, we need to show that for every rotation R, there is another rotation S such that $S \cdot R$ is no rotation at all If the angle for R is θ, then the angle $(2\pi - \theta)$ will be the angle for a rotation S which will return the circle to its original position.

Exercise 5. To show that the set of integers, with the binary operation of multiplication, is a group we have to show closure, associativity, the existence of an identity, and the existence of an inverse. Closure means that for any two integers, a and b, $a \times b$ is an integer. That is obvious. For any three integers a, b and c, associativity means $(a \times b) \times c = a \times (b \times c)$; this also is one of the properties of multiplication of the integers. And 1 is the identity element for multiplication of integers. What about the existence of an inverse? The inverse of multiplying by an integer a is to divide by a, or equivalently, to multiply by $1/a$. But $1/a$ will not be an integer, except for $a = 1$.

Exercise 6. Consider Big Ben; Z_{12} is also known as clock arithmetic.

Exercise 7. This one takes a little effort! The total amplitude, for the interfering alternatives of left and right slits is

$$A(x) = A_L(x) + A_R(x) = e^{i\kappa\left(\sqrt{d^2+\delta^2}+\sqrt{\ell^2+(x-\delta)^2}\right)}$$
$$+ e^{i\kappa\left(\sqrt{d^2+\delta^2}+\sqrt{\ell^2+(x+\delta)^2}\right)}$$
$$= e^{i\kappa\sqrt{d^2+\delta^2}}\left[e^{i\kappa\sqrt{\ell^2+(x-\delta)^2}} + e^{i\kappa\sqrt{\ell^2+(x+\delta)^2}}\right]$$

According to the central procedure, we now take this amplitude and multiply it by its conjugate in order to get the probability, $P(x) = |A(x)|^2$, give or take the proportionality constant Φ. We will just forget about Φ and pretend that it is 1; that is fine here. Then $P(x) = A^*(x)A(x)$ is

$$e^{-i\kappa\sqrt{d^2+\delta^2}}e^{i\kappa\sqrt{d^2+\delta^2}}\left[e^{i\kappa\sqrt{\ell^2+(x-\delta)^2}} + e^{i\kappa\sqrt{\ell^2+(x+\delta)^2}}\right]^*$$

$$\times \left[e^{i\kappa\sqrt{\ell^2+(x-\delta)^2}} + e^{i\kappa\sqrt{\ell^2+(x+\delta)^2}}\right]$$

$$= \left[e^{i\kappa\sqrt{\ell^2+(x-\delta)^2}} + e^{i\kappa\sqrt{\ell^2+(x+\delta)^2}}\right]^* \left[e^{i\kappa\sqrt{\ell^2+(x-\delta)^2}} + e^{i\kappa\sqrt{\ell^2+(x+\delta)^2}}\right]$$

$$= \left|e^{i\kappa\sqrt{\ell^2+(x-\delta)^2}}\right|^2 + \left|e^{i\kappa\sqrt{\ell^2+(x+\delta)^2}}\right|^2$$

$$+ 2\Re\left[e^{i\kappa\sqrt{\ell^2+(x-\delta)^2}-i\kappa\sqrt{\ell^2+(x+\delta)^2}}\right]$$

$$= 2 + 2\Re\left[e^{i\kappa\sqrt{\ell^2+(x-\delta)^2}-i\kappa\sqrt{\ell^2+(x+\delta)^2}}\right].$$

When ε is much smaller than 1, $\sqrt{1-\varepsilon^2}$ is very close to $1 - \varepsilon/2$. Because both δ and x are much smaller than ℓ, the exponent is

$$i\kappa\sqrt{\ell^2 + (x-\delta)^2} - i\kappa\sqrt{\ell^2 + (x+\delta)^2}$$

$$= i\kappa\ell\left[\sqrt{1 + \frac{(x-\delta)^2}{\ell^2}} - \sqrt{1 + \frac{(x+\delta)^2}{\ell^2}}\right]$$

$$\cong i\kappa\ell\left[\left(1 + \frac{(x-\delta)^2}{2\ell^2}\right) - \left(1 + \frac{(x+\delta)^2}{2\ell^2}\right)\right]$$

$$= \left(\frac{i\kappa\ell}{2\ell^2}\right)\left[(x^2 - 2x\delta + \delta^2) - (x^2 + 2x\delta + \delta^2)\right]$$

$$= -\left(\frac{2i\kappa\delta x}{\ell}\right).$$

As a result,

$$P(x) \propto 2 + 2\Re\left[e^{i\kappa\sqrt{\ell^2+(x-\delta)^2}-i\kappa\sqrt{\ell^2+(x+\delta)^2}}\right]$$

$$= 2 + 2\Re\left[e^{-(2i\kappa\delta/\ell)x}\right]$$

$$= 2(1 + \cos((2\kappa\delta/\ell)x)),$$

and so

$$P(x) \propto \cos^2(\kappa \delta x / \ell).$$

Exercise 8. The total amplitude for just the left slit is

$$A(x) = A_L(x) = e^{i\kappa\left(\sqrt{d^2+\delta^2} + \sqrt{\ell^2+(x-\delta)^2}\right)}$$

$$= e^{i\kappa\sqrt{d^2+\delta^2}} \left[e^{i\kappa\sqrt{\ell^2+(x-\delta)^2}} \right].$$

Then $P(x) = A^*(x)A(x)$ is

$$e^{-i\kappa\sqrt{d^2+\delta^2}} e^{i\kappa\sqrt{d^2+\delta^2}} \left[e^{i\kappa\sqrt{\ell^2+(x-\delta)^2}} \right]^* \left[e^{i\kappa\sqrt{\ell^2+(x-\delta)^2}} \right]$$

$$= \left[e^{i\kappa\sqrt{\ell^2+(x-\delta)^2}} \right]^* \left[e^{i\kappa\sqrt{\ell^2+(x-\delta)^2}} \right]$$

$$= \left| e^{i\kappa\sqrt{\ell^2+(x-\delta)^2}} \right|^2 = 1.$$

Exercise 9. Start by writing the amplitudes in polar form: $A_1 = R_1 e^{i\phi_1}$ and $A_2 = R_2 e^{i\phi_2}$. The probability is

$$P \propto |A_1 + A_2|^2 = |R_1 e^{i\phi_1} + R_2 e^{i\phi_2}|^2$$

$$= R_1^2 + R_2^2 + 2R_1 R_2 \left(e^{i(\phi_1-\phi_2)} + e^{-i(\phi_1-\phi_2)} \right).$$

Multiply by an element $e^{i\theta}$ of the circle group $U(1)$ and recalculate P.

$$P \propto |R_1 e^{i\phi_1+\theta} + R_2 e^{i\phi_2+\theta}|^2$$

$$= R_1^2 + R_2^2 + 2R_1 R_2 \left(e^{i(\phi_1+\theta-\phi_2-\theta)} + e^{-i(\phi_1+\theta-\phi_2-\theta)} \right)$$

$$= R_1^2 + R_2^2 + 2R_1 R_2 \left(e^{i(\phi_1-\phi_2)} + e^{-i(\phi_1-\phi_2)} \right).$$

Observe that θ disappears from the final expression.

Exercise 10. From $E = p^2/2m$, substitute $p = 2\pi\hbar/\lambda$ to get $E = (2\pi\hbar)^2/(2m\lambda^2)$. From Figure 13.1, the wavelengths are $2L$, L, $2L/3$, etc. Therefore, $\lambda = 2L/n$, $n = 1, 2, 3, \ldots$ and this may be substituted into $E = (2\pi\hbar)^2/(2m\lambda^2)$ to get $E_n = n^2(\pi\hbar)^2/(2mL^2)$. But $L = \pi \times 10^{-9}$ m; substitute this in for the final result.

Exercise 11. Start from Equation (19),

$$P_{+\hbar/2} = |A_{+\hbar/2}|^2 = \frac{\cos\theta + 1}{2}.$$

Write the amplitude in polar form: $A_{+\hbar/2} = Re^{i\xi}$. The phase factor ξ will disappear in forming $|A_{+\hbar/2}|^2$ and cannot be determined. By convention, it is zero. It is only possible to find the magnitude of the amplitude, R, to measure $+\hbar/2$.

$$R^2 = \frac{\cos\theta + 1}{2} = \cos^2\frac{\theta}{2}.$$

For the case of measuring $-\hbar/2$, the equivalent of Equation (19) is

$$P_{-\hbar/2} = |A_{-\hbar/2}|^2 = \frac{1 - \cos\theta}{2},$$

so that in this case, $A_{-\hbar/2} = \sin(\theta/2)e^{i\xi}$. The phase factor ξ again cannot be determined; in this case, the convention is that $\xi = \pi$, i.e. that $A_{-\hbar/2} = -\sin(\theta/2)$.

Exercise 12. A tough one! But that is why all the answers are in the back for you.

Since any combination of three characteristics G, S, and H has a probability, and probabilities are positive real numbers, the sum of two such probabilities is also zero or more. So

$$P(G, \neg S, H) + P(\neg G, S, \neg H) \geq 0.$$

Because these properties either are or are not (e.g., the sock either is or is not striped),

$$P(G, \neg S, \neg H) + P(G, S, \neg H) = P(G, \neg H).$$

That is, the probability of a green, not striped, hole-free sock, added to the probability of a green, striped, hole free sock is the probability of a green, hole-free sock.

Now add these two equations.

$$P(G, \neg S, H) + P(\neg G, S, \neg H) + P(G, \neg S, \neg H)$$
$$+ P(G, S, \neg H) \geq P(G, \neg H).$$

Because a sock either does or does not have a hole, $P(G, \neg S, H) + P(G, \neg S, \neg H) = P(G, \neg S)$. The probability that the sock is green, not striped and has holes, $P(G, \neg S, H)$, plus the probability that the sock is green, not striped and does not have holes, $P(G, \neg S, \neg H)$ is the probability that the sock is green and not striped, $P(G, \neg S)$:

$$P(G, \neg S) + P(\neg G, S, \neg H) + P(G, S, \neg H) \geq P(G, \neg H).$$

Similarly, $P(\neg G, S, \neg H) + P(G, S, \neg H) = P(S, \neg H)$:

$$P(G, \neg S) + P(S, \neg H) \geq P(G, \neg H).$$

This proves the theorem.

Exercise 13. Equation (22) is

$$s^2/4 + (t - s)^2/4 \geq t^2/4.$$

Multiply both sides by 4 and subtract s^2 from both sides:

$$(t - s)^2 \geq t^2 - s^2.$$

Use $(t - s)^2 = t^2 - 2st + s^2$, and subtract s^2 a second time:

$$t^2 - 2st \geq t^2 - 2s^2.$$

Subtract t^2 and divide by -2, changing of course $\geq$ to $\leq$:

$$st \leq s^2.$$

Divide by s, which is a positive number:

$$t \leq s.$$

Exercise 14. For a vector in Minkowski space, $s = 0$ means $(x_A - x_B)^2 = c^2(t_A - t_B)^2$, which is to say $(x_A - x_B) = \pm c(t_A - t_B)$. Divide both sides by $(t_A - t_B)$ and get

$$(x_A - x_B)/(t_A - t_B) = \pm c.$$

So $s = 0$ is the case for a particle traveling (either in the positive or negative direction) at the speed of light. On Figure 22.1, that would be a vector pointing at either 45° or 135° to the $+x$ axis.

Exercise 15. If you are wearing one of those T-shirts that reads "Another day and I didn't use algebra"... take it off. Then

$$s'^2 = (x'_A - x'_B)^2 + (y'_A - y'_B)^2$$

$$= (\cos(\phi)x_A + \sin(\phi)y_A - \cos(\phi)x_B - \sin(\phi)y_B)^2$$

$$\quad + (-\sin(\phi)x_A + \cos(\phi)y_A + \sin(\phi)x_B - \cos(\phi)y_B)^2$$

$$= \cos^2(\phi)x_A^2 + \sin^2(\phi)y_A^2 + \cos^2(\phi)x_B^2 + \sin^2(\phi)y_B^2$$

$$\quad + 2(\cos(\phi)\sin(\phi)x_A y_A - \cos^2(\phi)x_A x_A - \cos(\phi)\sin(\phi)x_A y_B)$$

$$\quad + 2(-\cos(\phi)\sin(\phi)y_A x_B - \sin^2(\phi)y_A y_B + \cos(\phi)\sin(\phi)x_B y_B)$$

$$\quad + \sin^2(\phi)x_A^2 + \cos^2(\phi)y_A^2 + \sin^2(\phi)x_B^2 + \cos^2(\phi)y_B^2$$

$$\quad + 2(-\sin(\phi)\cos(\phi)x_A y_A - \sin^2(\phi)x_A x_B + \sin(\phi)\cos(\phi)x_A y_B)$$

$$\quad + 2(\cos(\phi)\sin(\phi)y_A x_B - \cos^2(\phi)y_A y_B - \sin(\phi)\cos(\phi)x_B y_B)$$

$$= \cos^2(\phi)x_A^2 + \sin^2(\phi)y_A^2 + \cos^2(\phi)x_B^2 + \sin^2(\phi)y_B^2$$

$$\quad + 2(-\cos^2(\phi)x_A x_A) + 2(-\sin^2(\phi)y_A y_B)$$

$$\quad + \sin^2(\phi)x_A^2 + \cos^2(\phi)y_A^2 + \sin^2(\phi)x_B^2 + \cos^2(\phi)y_B^2$$

$$\quad + 2(-\sin^2(\phi)x_A x_B) + 2(-\cos^2(\phi)y_A y_B)$$

$$= \cos^2(\phi)x_A^2 + \sin^2(\phi)y_A^2 + \cos^2(\phi)x_B^2 + \sin^2(\phi)y_B^2$$

$$\quad - 2(x_A x_A + y_A y_B)$$

$$\quad + \sin^2(\phi)x_A^2 + \cos^2(\phi)y_A^2 + \sin^2(\phi)x_B^2 + \cos^2(\phi)y_B^2$$

$$= x_A^2 + y_A^2 + x_B^2 + y_B^2 - 2(x_A x_A + y_A y_B)$$

$$= x_A^2 - 2x_A x_A + y_A^2 + x_B^2 - 2y_A y_B + y_B^2$$

$$= (x_A - y_A)^2 + (x_B - y_B)^2,$$

obviously.

Exercise 16. Define the three matrices

$$A = \begin{bmatrix} a_{11} & \cdots & a_{1n} \\ \vdots & \ddots & \vdots \\ a_{n1} & \cdots & a_{nn} \end{bmatrix} \quad B = \begin{bmatrix} b_{11} & \cdots & b_{1n} \\ \vdots & \ddots & \vdots \\ b_{n1} & \cdots & b_{nn} \end{bmatrix}$$

$$C = \begin{bmatrix} c_{11} & \cdots & c_{1n} \\ \vdots & \ddots & \vdots \\ c_{n1} & \cdots & c_{nn} \end{bmatrix}$$

The task is to prove that $(AB)C$ is the same as $A(BC)$. The first step is to multiply A by B,

$$(AB) = \begin{bmatrix} \sum_{i=1}^{n} a_{1i}b_{i1} & \cdots & \sum_{i=1}^{n} a_{1i}b_{in} \\ \vdots & \ddots & \vdots \\ \sum_{i=1}^{n} a_{ni}b_{i1} & \cdots & \sum_{i=1}^{n} a_{ni}b_{in} \end{bmatrix}$$

and then this by C, giving

$$(AB)C = \begin{bmatrix} \sum_{i=1}^{n} a_{1i}b_{i1} & \cdots & \sum_{i=1}^{n} a_{1i}b_{in} \\ \vdots & \ddots & \vdots \\ \sum_{i=1}^{n} a_{ni}b_{i1} & \cdots & \sum_{i=1}^{n} a_{ni}b_{in} \end{bmatrix} \begin{bmatrix} c_{11} & \cdots & c_{1n} \\ \vdots & \ddots & \vdots \\ c_{n1} & \cdots & c_{nn} \end{bmatrix}$$

$$= \begin{bmatrix} \sum_{j=1}^{n} \left(\sum_{i=1}^{n} a_{1i}b_{ij} \right) c_{j1} & \cdots & \sum_{j=1}^{n} \left(\sum_{i=1}^{n} a_{1i}b_{ij} \right) c_{jn} \\ \vdots & \ddots & \vdots \\ \sum_{j=1}^{n} \left(\sum_{i=1}^{n} a_{ni}b_{ij} \right) c_{j1} & \cdots & \sum_{j=1}^{n} \left(\sum_{i=1}^{n} a_{ni}b_{ij} \right) c_{jn} \end{bmatrix}$$

$$= \sum_{i,j=1}^{n} \begin{bmatrix} a_{1i}b_{ij}c_{j1} & \cdots & a_{1i}b_{ij}c_{jn} \\ \vdots & \ddots & \vdots \\ a_{ni}b_{ij}c_{j1} & \cdots & a_{ni}b_{ij}c_{jn} \end{bmatrix}$$

That might look awful, but all that goes into it is the definition of matrix multiplication, applied twice.

Similarly, A(BC) is

$$
A(BC) =
\begin{bmatrix}
a_{11} & \cdots & a_{1n} \\
\vdots & \ddots & \vdots \\
a_{n1} & \cdots & a_{nn}
\end{bmatrix}
\begin{bmatrix}
\sum_{i=1}^{n} b_{1i}c_{i1} & \cdots & \sum_{i=1}^{n} b_{1i}c_{in} \\
\vdots & \ddots & \vdots \\
\sum_{i=1}^{n} b_{ni}c_{i1} & \cdots & \sum_{i=1}^{n} b_{ni}c_{in}
\end{bmatrix}
$$

$$
=
\begin{bmatrix}
\sum_{j=1}^{n} a_{1j}\left(\sum_{i=1}^{n} b_{ji}c_{i1}\right) & \cdots & \sum_{j=1}^{n} a_{1j}\left(\sum_{i=1}^{n} b_{ji}c_{in}\right) \\
\vdots & \ddots & \vdots \\
\sum_{j=1}^{n} a_{nj}\left(\sum_{i=1}^{n} b_{ji}c_{i1}\right) & \cdots & \sum_{j=1}^{n} a_{nj}\left(\sum_{i=1}^{n} b_{ji}c_{in}\right)
\end{bmatrix}
$$

$$
= \sum_{i,j=1}^{n}
\begin{bmatrix}
a_{1j}b_{ji}c_{i1} & \cdots & a_{1j}b_{ji}c_{in} \\
\vdots & \ddots & \vdots \\
a_{nj}b_{ji}c_{i1} & \cdots & a_{nj}b_{ji}c_{in}
\end{bmatrix}
$$

But, as the sums over i and j go over the same range, 1 to n, it is permissible to relabel the indices i and j:

$$
A(BC) = \sum_{j,i=1}^{n}
\begin{bmatrix}
a_{1i}b_{ij}c_{j1} & \cdots & a_{1i}b_{ij}c_{jn} \\
\vdots & \ddots & \vdots \\
a_{ni}b_{ij}c_{j1} & \cdots & a_{ji}b_{ij}c_{jn}
\end{bmatrix}.
$$

This is the same as the previous result, so $(AB)C = A(BC)$.

Exercise 17. Looking at Equations (28) and (32), the determinant is

$$
ad - bc = \cos^2(\phi) + \sin^2(\phi) = 1
$$

and 1 is not 0.

Exercise 18. If the perpendicular is drawn at ct_0 on the ct axis, the length between the ct and ct' axes will be vt_0. Then you only need the definition of $\tan(\phi)$.

Exercise 19. If you got this one without help, you deserve some ice cream. And not any shabby low-fat kind of ice cream, either!

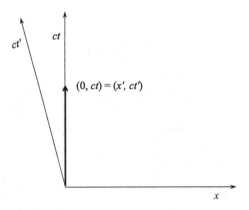

Figure 42.1: A particle at rest in the (x, ct) frame.

Figure 42.1 is the same as Figure 24.1, but with a different vector. This vector shows the motion in Minkowski space of an object that is not moving in the unprimed coordinate system. In the boosted system, this object has a steadily increasing x'; the object is moving in the positive x' direction.

Because the length of the vector is the same in the two coordinate systems,

$$-(ct)^2 = (x')^2 - (ct')^2.$$

The velocity in the boosted, primed, coordinates is $v = x'/t'$, so

$$-(ct)^2 = v^2(t')^2 - (ct')^2.$$

Divide by $-c^2$, substitute in β, factor out t', divide by $(1 - \beta^2)$ and then substitute in γ to get

$$\gamma^2(t)^2 = (t')^2.$$

Take the square root on both sides, to get

$$t' = \pm\gamma t.$$

However, γ is positive, as are both t and t', so the right sign of the square root is the positive one. Now we know the lower right element of the matrix:

$$\begin{bmatrix} x' \\ ct' \end{bmatrix} = \begin{bmatrix} U & V \\ W & \gamma \end{bmatrix} \begin{bmatrix} x \\ ct \end{bmatrix}.$$

where U, V, and W are still unknown matrix elements. They can be found by applying our symmetry principle $(s)^2 = (s')^2$:

$$x^2 - (ct)^2 = x'^2 - (ct')^2$$

$$= (Ux + Vct)^2 - (Wx + \gamma ct)^2$$

$$= (U^2 - W^2)x^2 + 2(UV - W\gamma)xct + (V^2 - \gamma^2)c^2t^2$$

Because this equation must hold when $x = 0$ as shown in Figure 42.1, we have $-1 = V^2 - \gamma^2$, so $V = \pm\beta\gamma$. This equation must also hold when $ct = 0$, and therefore, $1 = (U^2 - W^2)$. Similarly, matching the xct terms when $x = ct$ gives us $0 = UV - W\gamma$.

Substitute $V = \pm\beta\gamma$ into $0 = UV - W\gamma$, and get $W = \pm\beta U$, where if $V = +\beta\gamma$ then $W = +\beta U$, but if $V = -\beta\gamma$ then $W = -\beta U$. Next, substitute $W = \pm\beta U$ into $1 = (U^2 - W^2)$ and get $1 = (1 - \beta^2)U^2$ which means $U = \gamma$. Now our boost matrix is

$$\begin{bmatrix} x' \\ ct' \end{bmatrix} = \begin{bmatrix} \gamma & \pm\beta\gamma \\ W & \gamma \end{bmatrix} \begin{bmatrix} x \\ ct \end{bmatrix}.$$

But since $W = \pm\beta U$ and $U = \gamma$, $W = \pm\beta\gamma$. Now our matrix is

$$\begin{bmatrix} x' \\ ct' \end{bmatrix} = \begin{bmatrix} \gamma & \pm\beta\gamma \\ \pm\beta\gamma & \gamma \end{bmatrix} \begin{bmatrix} x \\ ct \end{bmatrix},$$

and the sign in the upper right, whether positive or negative, is the same as the sign in the lower left. So which sign is it?

Our original vector $(0, ct)$ in the unprimed frame of the figure has $x' > 0$ in the boosted coordinates. So

$$\begin{bmatrix} x' \\ ct' \end{bmatrix} = \begin{bmatrix} \gamma & \pm\beta\gamma \\ \pm\beta\gamma & \gamma \end{bmatrix} \begin{bmatrix} 0 \\ ct \end{bmatrix} = \begin{bmatrix} \pm\beta\gamma ct \\ \gamma ct \end{bmatrix},$$

and therefore, in order for x' to be positive, we need $+\beta\gamma$ in the corners.

Warning! The $+\beta\gamma$ is the right choice for a boost matrix that takes an object at rest and gives it a velocity v in the positive direction. That is the same as taking an object at rest and giving ourselves, as observers, a velocity in the negative direction. But if we give ourselves a velocity v in the positive direction, that is the same as giving the object a velocity in the negative direction and in that case, we would want the negative square root. So watch out for the signs in the corners!

Exercise 20. From Equation (33),

$$x' = \gamma x + \beta\gamma ct$$
$$ct' = \beta\gamma x + \gamma ct$$

so

$$
\begin{aligned}
(x')^2 - (ct')^2 &= (\gamma x + \beta\gamma ct)^2 - (\beta\gamma x + \gamma ct)^2 \\
&= \gamma^2\{(x^2 + 2x\beta ct + \beta^2 c^2 t^2) - (\beta^2 x^2 + 2\beta xct + c^2 t^2)\} \\
&= \gamma^2\{(x^2 - \beta^2 x^2) + \beta^2 c^2 t^2 - c^2 t^2)\} \\
&= \gamma^2\{(1 - \beta^2)x^2 - (1 - \beta^2)c^2 t^2\} \\
&= \{x^2 - c^2 t^2\}.
\end{aligned}
$$

Exercise 21. From Equations (32) and (34), the inverse of the boost matrix in Equation (34) is

$$
\begin{bmatrix} \gamma & \beta\gamma \\ \beta\gamma & \gamma \end{bmatrix}^{-1} = \frac{1}{\gamma^2 - \beta^2\gamma^2} \begin{bmatrix} \gamma & -\beta\gamma \\ -\beta\gamma & \gamma \end{bmatrix}
$$

but from the definition of γ,

$$
\frac{1}{\gamma^2 - \beta^2\gamma^2} = \left(\frac{1}{\gamma^2}\right)\frac{1}{1 - \beta^2} = \left(\frac{1}{\gamma^2}\right)\gamma^2 = 1.
$$

That proves that the determinant of the inverse of the boost matrix is 1. The determinant of the boost matrix itself is

$$
\det \begin{bmatrix} \gamma & \beta\gamma \\ \beta\gamma & \gamma \end{bmatrix} = \gamma^2 - \beta^2\gamma^2
$$

which is also 1. Returning to the main task, the inverse of the matrix in Equation (33) is therefore

$$\begin{bmatrix} \gamma & \beta\gamma \\ \beta\gamma & \gamma \end{bmatrix}^{-1} = \begin{bmatrix} \gamma & -\beta\gamma \\ -\beta\gamma & \gamma \end{bmatrix}$$

which is, in retrospect perhaps, obvious. The boost matrix corresponds to a velocity $v = c\beta$. To undo that, we want a matrix for a velocity $-v$, corresponding to a matrix with $-\beta$ in place of β.

Now apply this inverse to both sides of Equation (34):

$$\begin{bmatrix} \gamma & -\beta\gamma \\ -\beta\gamma & \gamma \end{bmatrix} \begin{bmatrix} 1 \\ 0 \end{bmatrix} = \begin{bmatrix} \gamma & \beta\gamma \\ \beta\gamma & \gamma \end{bmatrix}^{-1} \begin{bmatrix} \gamma & \beta\gamma \\ \beta\gamma & \gamma \end{bmatrix} \begin{bmatrix} x \\ ct \end{bmatrix}$$

to get

$$\begin{bmatrix} \gamma \\ -\beta\gamma \end{bmatrix} = \begin{bmatrix} x \\ ct \end{bmatrix}.$$

Exercise 22. Draw the ct' axis to be parallel to the vector AB. As the sign of the velocity has changed, Equation (35) becomes

$$\begin{bmatrix} \gamma \\ +\beta\gamma \end{bmatrix} = \begin{bmatrix} x \\ ct \end{bmatrix}$$

so Figure 24.2 becomes Figure 42.2.

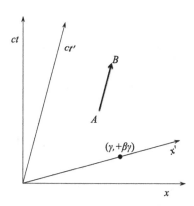

Figure 42.2: Figure 24.2, after changing the signs of the velocities.

Exercise 23. Start with the combined boost equation

$$\begin{bmatrix} \gamma_T & \beta_T\gamma_T \\ \beta_T\gamma_T & \gamma_T \end{bmatrix} = \begin{bmatrix} \gamma_2 & \beta_2\gamma_2 \\ \beta_2\gamma_2 & \gamma_2 \end{bmatrix} \begin{bmatrix} \gamma_1 & \beta_1\gamma_1 \\ \beta_1\gamma_1 & \gamma_1 \end{bmatrix}$$

and find the upper left and upper right elements. They are, from multiplying the two matrices on the right,

$$\gamma_T = \gamma_1\gamma_2(1 + \beta_1\beta_2)$$

and

$$\beta_T\gamma_T = \gamma_1\gamma_2(\beta_1 + \beta_2)$$

The ratio of these two equations gives Equation (38).

Exercise 24. The simplest way to do this is to graph the function on the computing device of your choice.

Exercise 25. For $\beta = 0.999651076$, $\gamma = 1/\sqrt{1 - \beta^2} = 1/\sqrt{1 - (0.999651076)^2} = 37.86$. Then the decay time becomes larger by a factor of 37.86, so instead of being $2\,\mu s$ it is $75.7\,\mu s$. Because β is so close to 1, the distance that the muon travels before decaying, $\beta\gamma c(2\,\mu s) \cong c(75.7\,\mu s) = 22.7\,km$. That means that a cosmic ray muon created in the upper atmosphere can easily reach the surface of the earth before decaying.

Exercise 26. B. May, *A Survey of Radial Velocities in the Zodiacal Dust Cloud*, PhD. diss., Imperial College of Science, Technology and Medicine, London, 2007.

Exercise 27. The binomial theorem is

$$(a + b)^n = a^n + na^{(n-1)}b + (n/2)(n - 1)a^{(n-2)}b^2$$
$$+ (n/6)(n - 1)(n - 2)a^{(n-3)}b^3 \ldots$$

Apply it for $a = 1$, $b = -\beta^2$ and $n = -1/2$. That gives an infinite series,

$$1/\sqrt{1 - \beta^2} \cong 1^{-1/2} + (-1/2)1^{-3/2}(-\beta^2)$$
$$+ ((-1/2)/2)(-3/2)1^{-5/2}(-\beta^2)^2 \ldots$$

where each successive term is a higher and higher power of $-\beta^2$. Because β is small, these later terms are negligible; the first two terms are all we need to keep, and they are

$$1/\sqrt{1-\beta^2} \cong 1^{-\frac{1}{2}} + (-1/2)1^{-\frac{3}{2}}(-\beta^2) = 1 + \beta^2/2.$$

Exercise 28. Equation (52) was

$$p^{(M)} = \gamma m \begin{bmatrix} v \\ c \end{bmatrix} = \begin{bmatrix} p \\ E/c \end{bmatrix},$$

which is to say $p = m\gamma v$, and $E/c = m\gamma c$. Divide p by E/c:

$$\beta = \frac{v}{c} = cp/E.$$

Then, in Equation (54),

$$E^2 = \left(\frac{cp}{\beta}\right)^2 = (mc^2)^2 + (cp)^2$$

which, for a particle at the speed of light ($\beta = 1$) is

$$(cp)^2 = (mc^2)^2 + (cp)^2.$$

For a particle of nonzero momentum, $p > 0$, this can only be true if the first term, $(mc^2)^2$ is zero, which in turn requires m to be zero.

Exercise 29. Starting at

$$a^*a + b^*b = (u_{11}a + u_{12}b)^*(u_{11}a + u_{12}b) + (u_{21}a + u_{22}b)^*(u_{21}a + u_{22}b),$$

expand the right hand side to

$$(u_{11}^*u_{11} + u_{21}^*u_{21})(a^*a) + (u_{11}^*u_{12} + u_{21}^*u_{22})(a^*b)$$

$$+(u_{12}^*u_{11} + u_{22}^*u_{21})(ab^*) + (u_{12}^*u_{12} + u_{22}^*u_{22})(b^*b)$$

and equate the corresponding terms in a^*a, a^*b, b^*a, and b^*b.

Exercise 30. If $U = \begin{pmatrix} u_{11} & u_{12} \\ u_{21} & u_{22} \end{pmatrix}$, then $U^\dagger = \begin{pmatrix} u_{11}^* & u_{21}^* \\ u_{12}^* & u_{22}^* \end{pmatrix}$, from the definition of the Hermitian conjugate. Then write out the product

$$\begin{pmatrix} u_{11}^* & u_{21}^* \\ u_{12}^* & u_{22}^* \end{pmatrix} \begin{pmatrix} u_{11} & u_{12} \\ u_{21} & u_{22} \end{pmatrix}.$$

Exercise 31. For σ_1 to be in $SU(2)$, Equation (66) says that it must be of the form

$$U = \begin{bmatrix} a & b \\ -b^* & a^* \end{bmatrix}$$

but σ_1 cannot be written in this form, because that would require that the lower left corner, which is 1, be the negative of the conjugate of upper right corner, i.e., $1 = -1^*$, which is not true. Similarly, for σ_2 to be in $SU(2)$, we would need $-(-i)^* = -i$ in the lower left corner, but actually there is i there. For σ_3, we would need $(1)^* = 1$ in the lower right corner, but actually there is -1 there.

Exercise 32. This "lowering operator" as it is also known, is

$$\sigma^- = (\sigma_1 - i\sigma_2)/2 = \frac{1}{2}\left\{ \begin{bmatrix} 0 & 1 \\ 1 & 0 \end{bmatrix} - i \begin{bmatrix} 0 & -i \\ i & 0 \end{bmatrix} \right\}$$

$$= \frac{1}{2}\left\{ \begin{bmatrix} 0 & 1 \\ 1 & 0 \end{bmatrix} - \begin{bmatrix} 0 & 1 \\ -1 & 0 \end{bmatrix} \right\}$$

$$= \frac{1}{2} \begin{bmatrix} 0 & 0 \\ 2 & 0 \end{bmatrix} = \begin{bmatrix} 0 & 0 \\ 1 & 0 \end{bmatrix}.$$

When applied to a proton,

$$\sigma^- \begin{bmatrix} 1 \\ 0 \end{bmatrix} = \begin{bmatrix} 0 & 0 \\ 1 & 0 \end{bmatrix} \begin{bmatrix} 1 \\ 0 \end{bmatrix} = \begin{bmatrix} 0 \\ 1 \end{bmatrix},$$

which is a neutron.

Exercise 33. λ^+ applied to n is

$$\begin{bmatrix} 0 & 1 & 0 \\ 0 & 0 & 0 \\ 0 & 0 & 0 \end{bmatrix} \begin{bmatrix} 0 \\ 1 \\ 0 \end{bmatrix} = \begin{bmatrix} 1 \\ 0 \\ 0 \end{bmatrix},$$

a proton. λ^- applied to p is

$$\begin{bmatrix} 0 & 0 & 0 \\ 1 & 0 & 0 \\ 0 & 0 & 0 \end{bmatrix} \begin{bmatrix} 1 \\ 0 \\ 0 \end{bmatrix} = \begin{bmatrix} 0 \\ 1 \\ 0 \end{bmatrix},$$

a neutron.

Exercise 34. From Table 37.1, a Δ^{++} has charge $+2$ and 3 u quarks, so the charge of the u quark is $+2/3$. A Δ^- has charge -1 and $3d$ quarks, so the charge of the d quark is $-1/3$. An Ω^- also has charge -1 and $3s$ quarks, so the charge of the s quark is also $-1/3$.

Exercise 35. Again from Table 37.1, both the Δ^{++} and the Δ^- have strangeness 0, and contain respectively $3u$ and $3d$ quarks, so both the u and d quarks have strangeness 0. The Ω^- has strangeness -3 and has $3s$ quarks, so the strangeness of the s quark is -1.

Exercise 36. Assume that $\langle \vec{x}|\psi_{21}\rangle = Ae^{i\delta}\langle \vec{x}|\psi_{12}\rangle$, where the magnitude A is a positive real number. Since $P_{21} \propto |\langle \vec{x}|\psi_{21}\rangle|^2$,

$$P_{21} = \Phi|\langle \vec{x}|\psi_{21}\rangle|^2 = \Phi|Ae^{i\delta}\langle \vec{x}|\psi_{12}\rangle|^2$$

where Φ is the constant of proportionality from Equation (9). And

$$P_{12} = \Phi|\langle \vec{x}|\psi_{12}\rangle|^2,$$

so

$$P_{21} = \Phi|\langle \vec{x}|\psi_{21}\rangle|^2 = \Phi A^2|\langle \vec{x}|\psi_{12}\rangle|^2 = A^2 P_{12}.$$

But the probabilities P_{12} and P_{12} are equal because the two particles are indistinguishable, so their labels are arbitrary. Consequently, the real number $A^2 = 1$; and since the assumption is that A is a positive real number, it must be 1, and so the only possibility is that $\langle \vec{x}|\psi_{21}\rangle = e^{i\delta}\langle \vec{x}|\psi_{12}\rangle$.

Index

1-form, 129

antimatter, 14, 33, 127

baryons, 140
boost matrix, 103
bosons, 80, 168

cobalt
    decay, 11
color charge, 124, 155, 161
contravariant vs. covariant, 129

Dirac equation, 124

Eightfold Way, 149, 151
exclusive alternatives, 33, 35
experimental science
    fundamental law, 76

fermions, 80, 152, 168
Feynman diagram, 44
forces
    electromagnetic, 11, 55, 115
    gravity, 1, 27, 113, 115, 124
    strong, 11, 55, 124, 153, 155,
        157, 169
    weak, 11, 33, 45, 55, 125, 159
Fourier transform, 50
frames of reference, 101

γ and β, 103
gauge invariance, 137, 168
Gell-Mann, 149, 150
    matrices, 145
generating sets, 23
generator, 23
groups
    associative, 19
    circle, 20, 53, 99, 159
    closure, 19
    cyclic, 20, 21
    D4, 20
    matrix, 100
    Z_2, 20

hats
    conservation of, 72
hidden variables, 77
Higgs, 41, 55
    field, 161

interfering alternatives, 33, 35, 39
    indistinguishability, 31, 34,
        167
isospin, 127

kaon, 147
Klein-Gordon equation, 123

Lagrangian, 25, 35
lambda, 143

leptons, 140
Lorentz contraction, 111

matrix
    determinant, 100
matter waves, 63
mesons, 140, 147
    η, η?, 147
metaphysics, 66
models vs. theory, 128
muon, 33, 44, 109, 139

Ne'eman, 149
neutrino, 11, 14, 31, 109
nucleon, 127

parity, 9, 20
parton, 155
pion, 140
probability
    Bayesian v. frequentist, 52, 71, 87
projection function, 128, 132

quantum computing, 137
quark, 140, 150

Rabi
    Isidor, 139, 160
relativity
    general, 1, 94, 124
    principle of, 27
    special, 93, 113

spin, 12, 79
SU(n)
    recapped, 153, 160
symmetry
    broken, 2
    CP, 14, 125
    gauge invariance, 41
    terminology, 2

time dilation, 110, 111

unitary matrix, 132

wavefunction, 50
weak isospin, 141

Zweig, 149, 150

Printed in the United States
by Baker & Taylor Publisher Services